# THE ABONGO ABROAD

**THE COLD WAR IN GLOBAL PERSPECTIVE**
aims to publish innovative books that combine methods of political and cultural history; utilize previously untapped materials from foreign archives; and make significant contributions to the historiography of the social and cultural aspects of the Cold War.

PHILIP E. MUEHLENBECK, SERIES EDITOR

*The Abongo Abroad: Military-Sponsored Travel in Ghana, the United States, and the World, 1959–1992, by John V. Clune (inaugural volume, 2017)*

# THE AB★NGO ABROAD

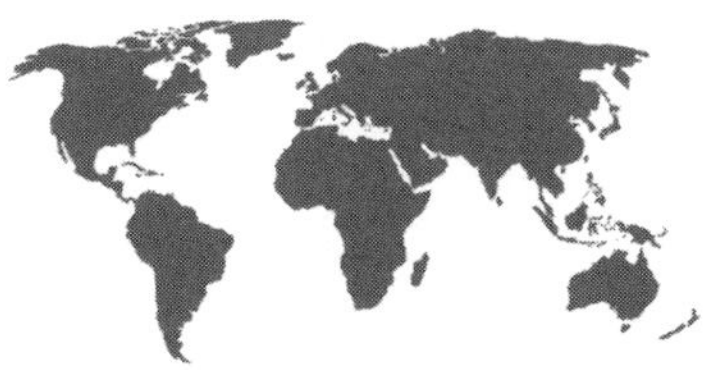

Military-Sponsored
Travel in Ghana,
the United States,
and the World,
1959–1992

**JOHN V. CLUNE**

Vanderbilt University Press | Nashville

First printing 2017

This book is printed on acid-free paper.
Manufactured in the United States of America

Library of Congress Cataloging-in-Publication Data
LC control number: 2016042796
LC classification number: UA695.G4 .C58 2017
Dewey classification number: 355.09667—dc23
LC record available at *lccn.loc.gov/2016042796*

ISBN 978-0-8265-2151-4 (hardcover)
ISBN 978-0-8265-2153-8 (ebook)

# Contents

# Acknowledgments

As all historians know, works of history are community efforts. Thanks go to Phil Muehlenbeck, who invited me to share his vision for a Social History of the Cold War, and Eli Bortz, who shepherded me through the process. I thank my colleagues and professional advisors, including Anton Rosenthal, Elizabeth MacGonagle, Marie Brown, Brent Steele, Holly Goerdel, and especially Sheyda Jahanbani, for carefully guiding this project in reasonable ways without dampening my enthusiasm. At the University of Kansas, Beverly Mack, Theodore Wilson, Jeffrey Moran, Jennifer Weber, Yacine Daddi Addoun, and Robert Schwaller encouraged me as a scholar and friend more than they know. Colleagues Edward Woodfin, Edward Kaplan, Grant Weller, and Mark Grotelueschen generously offered their timely, sage advice. Librarians Carmen Orth-Alfie and Sara Morris provided rapid assistance whenever I called. Special thanks to Alan Clune and Amanda Schlumpberger for their countless hours editing and clarifying my work.

Thanks to the US Air Force Academy's Department of History, especially Mark Wells, for sponsoring the fellowship that began this project, and the Air Force Institute of Technology's Civilian Institution section for administrative support while researching. The views expressed in this book are my own and do not reflect the official policy or position of the United States Air Force, the Department of Defense, or the US government. I thank the US Air Force Academy's Directorate of Education, which sponsored my research in Ghana, and Brian Bartee, the US Air Force attaché who coordinated my meeting with his Ghanaian counterpart, Derrick Attachie. Lieutenant Colonel Attachie arranged interviews with former Ghanaian officers, provided transportation to the Ghana Armed Forces Military Academy and Training Schools, the Command and Staff College, and introduced me to the Ghana Armed Forces Director of Public Relations, M'Bawine Atintande. Colonel Atintande provided a workspace and gave me free run of his library while his deputy Eric Aggrey-Quashie scoured the Directorate of Public Relations to create the most complete collection of peacekeeping journals and papers from *Ghana Armed Forces News* ever assembled for my research. At Ghana's Public Records and Archives Administration, Josiah Okyere's attentive assistance helped me navigate collections to make my time there productive and exhilarating. Thanks to the University of Kansas's College of Humanities for sponsoring my research at the United Nations archives and to the United Nations Archives and Record Management Section's amazingly attentive and patient archival staff. Finally, special thanks to my wife Jessica and children, for giving me a quiet space and a reason to work. It is to them and in honor of the one hundred Ghanaian soldiers who gave their lives as United Nations peacekeepers that I dedicate this work.

# Foreword

On a panel commemorating the fiftieth anniversary of the State Department's African Bureau at the 2009 SHAFR Annual Meeting, Thomas Noer, who may rightly be called the dean of US-Africa diplomatic historians, recalled a SHAFR meeting in the 1980s at which attendees were asked to split up into discussion groups based upon the geographical focus of their research interests. Noer was the lone scholar at the meeting whose primary focus was Africa. Two decades later, little had changed when I first began my graduate studies in the early 2000s. At the 2004 International Graduate Student Conference on the Cold War, I presented on a panel entitled, "The Cold War in Latin America and Africa." Out of the twenty papers given at the conference, I was the only scholar working on Africa, thus my paper was added to a panel that was otherwise exclusively focused on Latin America. Six years later, however, the field had changed dramatically. When I served as a panel chair and discussant at the same conference in 2010, seven of the conference's twenty-eight papers were related to Africa. In the early 2000s, following in the footsteps of path-blazing scholars such as Noer, Tim Borstelmann, and Richard Mahoney, Piero Gleijeses and Andy DeRoche became the first diplomatic historians to internationalize the study of US-African relations through the use of African archival research.[1] In recent years, a new generation of historians such as Ryan Irwin, Will Bishop, Jamie Miller, and myself are doing the same.[2] Now John Clune joins this club with this provocative study of the United States' military education and training program as it was experienced in both the United States and Ghana.

In *The Abongo Abroad*, John Clune adds several new aspects to the already flourishing scholarship on development and modernization theory. First, no previously published work does such an exemplary job explaining the Military Assistance Program and its relationship to the modernization and development policies of the Kennedy and Johnson administrations. Secondly, Clune's work demonstrates that, contrary to previous assumptions, the ideas and mindset which were the foundation of modernization theory lived on well past their presumed decline of the late 1960s by serving as the rationale for Military Internationalism in both the United States and Ghana into the late 1980s. Finally, this study is one of the few in the modernization theory genre that shows a non-US perspective.

This is the first intensive historical study of a program that deputy assistant secretary of defense for international security affairs Frank Kramer said "dollar for dollar it is the most important program we [the United States] run."[3] Designed to 'win the hearts and minds' of future military leaders of the developing world, the United States' Military Assistance Program (MAP), which began in 1949, and the later International Military Education and Training (IMET) (1979) continue to the present day. Yet during the Cold War, the United States government never did a systematic evaluation to measure the value of these programs for US national interests. Nor was there any appraisal of whether the programs promoted respect for human rights and civilian control of government and fostered pro-US feelings from the participants or if instead it nurtured the growth of foreign military dictators. Clune finds that "instead of questioning the apparent lack of fruit of several decades of military exchanges, American policy makers aggressively defended [IMET] programs as a precarious toehold of American influence" (165).

*The Abongo Abroad* is interesting in that it shows the Ghanaian side of the story, not only from the high-level governmental angle, but also from the level of individual soldiers to illustrate how international military travel could better the lives of individual soldiers and their families. In this context, Clune's idea of what he refers to as "institutionalized instability"—the "ongoing competition for training, education, and opportunities for promotion"—is a particularly valuable concept to help explain why the Ghanaian side was so keen to be the recipients of military travel opportunities from the United States, United Nations, and other sources.

It is this non-US angle that makes *The Abongo Abroad* such a good fit as the first publication in Vanderbilt University Press's new series, *The Cold War in Global Perspective.* This series aims to publish innovative books that combine methods of political and cultural history, utilize previously untapped materials from foreign archives, and make significant contributions to the historiography of the social and cultural aspects of the Cold War.

In *The Abongo Abroad*, Clune demonstrates an impressive mastery of the secondary literature in the areas of modernization theory, US military aid assistance, US-African relations, and African history more broadly. The primary source research is even more impressive, as many of the sources utilized in this study have never before been used in a historical study. In addition to using well-traversed material from the US National Archives, Clune also draws upon oral history interviews with a number of Ghanaians who took part in military-sponsored travel as well as an abundance of material from the national archives of Ghana, the United Nations, and local areas in both Ghana and the United States in which these programs were implemented.

Scholars of modernization theory, US military history, Ghanaian history, African history, and the social, cultural, and military history of the Cold War will all find this work a valuable addition to their bookshelf.

Philip E. Muehlenbeck
*Series Editor*

# THE ABONGO ABROAD

# Introduction

*A few ardent Muslims have been able to make their pilgrimage to Mecca, thus becoming* alhajis *while on "Op Sunrise." From time to time, a few of the officers and men with an interest in agriculture express a wish to visit some kibbutzim to acquaint themselves with Israeli farming techniques and know-how, which are well known to be among the best in the world. . . . In short, the Ghanbatt participation in "Op Sunrise" has been immensely beneficial. The troops have benefitted professionally by operating alongside so many other national contingents, and the travel to other parts of the world far from home is in itself a great education. Our officers and men went to the Pyramids of Giza, the sands of Sinai and the holy places they had heard of at Sunday School.*

– Lieutenant General Emmanuel Erskine,
*Mission with UNIFIL: An African Soldier's Reflections*

In *Cosmopolitanism: Ethics in a World of Strangers*, Ghanaian philosopher Kwame Anthony Appiah defends ordinary citizens on both the sending and receiving sides of the great projects of cultural imperialism in the twentieth century. They are not "blank slates on which global capitalism's moving finger writes its message, leaving behind another homogenized consumer as it moves on," he argues (111).[1] That attitude, he says, "is deeply condescending. And it isn't true." People distinguish the good from the bad that they see, and the lessons they take from "Westernization" are not necessarily the same ones "cultural imperialists" try to sell. Instead, Appiah reflects, "people in each place make their own uses for even the most famous global commodities" (113). Even more significant than the differences in power or wealth they reveal, exchange in global commodities also exposes participants' obligations to one another and provides opportunities for individuals to create new ties—"beyond," Appiah suggests, "those to whom we are related by the ties of kith and kind, or even the more formal ties of a shared citizenship" (iv). The global commodities of military education and training and international peacekeeping offered those opportunities and exposed those obligations.

This book searches international military educational exchanges and United Nations peacekeeping missions between the 1960s and 1980s for the uses people on both the sending and receiving ends had for the international experience. It argues that when the elites that formed American military assistance policy toward Ghana and many other newly independent states of the Global South in the decades after 1960, they expected the modernization of the state to result from their ideological intervention into individual Ghanaians' lives. American defense assistance planners expected that their training programs would make individual Ghanaian officers into modern men, who would form the Ghanaian armed forces into a modern institution within a modern nation. To American planners, modernity resembled the Western nation-state. Ghanaians often did live differently after participating in the global structure of military-sponsored international travel that included the United States' international military education and training program but that also encompassed training and education programs in other countries and included extensive participation in UN peacekeeping missions. Ghanaian individuals and families developed transnational friendships and reaped social and financial benefits from the exchange. But given the turmoil surrounding the Ghanaian state between 1960 and 1990, for Ghanaians, those opportunities often eroded national identities while creating new global citizens, within limits. Most importantly, the essential personal transformation was international; it could only be activated in an international setting, and its reference points were necessarily extra-national. For Ghanaian military families, modernity was international.

Several different groups interact in this account: American social scientists, diplomats, and foreign policy planners in the 1960s, 1970s, and 1980s, who tried to employ American military assistance to achieve American political objectives while delivering economic development to some other parts of the world; Ghanaian military-turned-political leaders who, in the three decades after independence in 1957, tried to leverage internationally available military training programs and Ghana's own moral authority as a pan-African leader for their political purposes; Ghanaian officers and their families in communities around military training institutions abroad; and Ghanaian soldiers of all ranks who served in UN peacekeeping operations in the Sinai between 1973 and 1979 and in Lebanon continuously after 1979 and their families, who bore the burdens of the separation and the danger and who obtained the financial benefits that came with UN peacekeeping service.

Travel, specifically military-sponsored international travel, connected the ideas behind international military education and peacekeeping policies to people in both Ghana and the United States between the early 1960s and 1980s. All participants in the global system of military-sponsored international travel approached the act with different ideas about what the travel signified, what opportunities it presented, and what change it intended to bring about. But both believed the travel inspired or revealed a new psychological orientation capable of transcending national boundaries, when necessary, and which I call Military Internationalism. This book explores the transactions that occurred over three decades between the ideologies and intents behind military assistance policy and the actual participants in that travel. Throughout the era that witnessed the rise of modernization ideology as the guiding principle of American foreign assistance policy in the 1960s, its supposed demise after the Vietnam War, and the Nixon, Ford, Carter, and Reagan eras, American military assistance policy to states like Ghana publicly proposed to create modern men and American allies at a low cost. When pressed to explain American assistance programs' apparent lack of success, however, policy makers routinely touted the *relationships* that these exchanges created as the highest good and the ultimate benefit of American aid. Internationalism, not nationalism, became the goal. Ghanaians, likewise, may have thought themselves more modern after training courses in the United States, India, the United Kingdom, or elsewhere or after a UN deployment that enabled them to purchase household goods not available in Ghana, but their version of modernity did not necessarily imply greater allegiance to their nation-state.

The American and Ghanaian governments made international education and training an essential aspect of their foreign policy in the 1960s, 1970s, and 1980s. Using archival sources from the US Department of State and US Department of Defense, published congressional and presidential documents, and modernization theorists' own work, Chapter 1 argues that in the United States, an alliance of academic elites and policy makers in the Kennedy administration blended the prior American assumption that the national military represented modernity in "developing societies" with their new emphasis on modernization as a Cold War strategy. The United States' international military education and training program became their preferred method to transform visiting elites from target states into "modern" men and American allies at a low cost. As an essential aspect of international education, the "Informational Program for Foreign Military Trainees and Visitors in the United States" made the American people themselves full participants and legitimate subjects for lessons in American-style modernity. Since that program made psychological transformation its primary purpose, not any specific or measurable change to the military capabilities or political structure of recipient states, American ideals about the exceptionalism of American society sustained the international military education and training program during the next three decades, even when most other aspects of American military assistance policy changed drastically.

Chapter 2 integrates archival sources describing the founding of Ghana's military academy and staff college and Ghana's own military-produced publications to argue that for Ghanaians after independence, just as for Americans, the social value of lessons in military leadership, management of modern state institutions, and the vague category of professionalism transcended the national location where those lessons occurred. Ghanaians consumed those lessons from a wide variety of educational and training facilities abroad in the 1960s and 1970s, but also built their own capacity to provide them to Ghanaians and military personnel from other African states as well. Meanwhile, the United States provided international military education grants as its only military assistance to Ghana during that time. The rise of Ghanaian Colonel Ignatius Acheampong to power in 1972 should have represented a crowning achievement of American military education, but instead he demonstrated its potential downside. The small American military education program was not only not enough to create "military professionalism" or secure a balanced civil-military relationship in another state, it may actually have encouraged discontented military leaders to seize power.

Chapter 3 explores how individual Ghanaians envisioned their role in international exchange. Travel conveyed tangible political and economic benefits, but this chapter also argues that individuals' outlooks became more transnational (or international, given the ongoing reliance on state structures) as a result. Ghanaian families employed the infrastructure of international military cooperation to form an alternate global identity that was not simply larger than the nation-state. Ghanaians in international military service represented some aspects of the modern Ghanaian nation—such as in Ghanaian participation in the United Nations or Ghanaian leadership among multilateral peacekeeping forces. But the idea of a Ghanaian nation was problematic even within Ghana during these decades, and Ghanaian families actively manipulated state structures for military travel in ways that skirted rather than reinforced nationalism. This chapter uses Ghanaian soldiers' memoirs, entries from peacekeeping journals in the Sinai and Lebanon in the 1970s and 1980s, and *Ghana Armed Forces News* newspapers to explore the way Ghanaian families employed international military travel as a transnational economic strategy. In that sense, Ghanaian military families abroad resembled other Ghanaians in the diaspora. For military wives, that service placed new expectations on them for work, family, and home, but it also provided them with the means to make claims against the state to fulfill them. Finally, the way individual Ghanaians embraced peace*making* on UN missions in Egypt and Lebanon also reveals that the global identity military service activated had imaginary roots accessible only through specifically military channels.

The final chapter argues that in the 1970s and 1980s, the intangible assumptions behind international military education and training—what I call Military Internationalism—did real work defending the assistance program as the US Congress questioned the role of American training in producing or encouraging dictators abroad. Department of Defense and State Department officials equated the program with the Fulbright scholarship and argued that the program's essential power to transform societies was through its humanity, not its militarism. They reiterated the long-held belief that life among Americans fundamentally changed people. That change was the first goal of the program, they argued, and society-wide improvement would result, but only in the long term. Careful rewording of the program to emphasize "internationally recognized human rights" and shifting justifications for American military assistance to Ghana in the years surrounding the Rawlings revolution exposed the international military education and training program as, essentially, an act of faith. But the original confidence that human-to-human interaction could transform visitors was not a myth. In American military assistance policy, it was often the only "real" thing.

* * *

Given the slowness of measurable results toward global peace, economic development across the Global South, or Ghanaian political development to appear between 1960 and 1990, international military travel did not yield the results Ghanaian or American policy makers expected. American defense policy makers and congressional critics expressed doubts about these international exchange programs repeatedly in the 1970s and 1980s but never abandoned them. Instead, they vaguely recognized that the essential historical process at work resisted measurement through statistics. International travel changed people's minds, and the world changed within the minds of people. What those involved saw differently about themselves, about their new international relationships, and about the world, is the subject of this work.

## The Case for Ghana

Unlike other examples of American military assistance in the Cold War that metastasized into full-blown proxy wars, created puppet governments that depended on military aid to maintain their political control, or transferred great quantities of American weapons over long periods, the case of Ghana demonstrates the intangible assumptions of Cold War military assistance exceptionally well.[2] During the Cold War, the United States and Soviet Union routinely employed military assistance as capital in their ideological proxy battles.[3] The way both US and Ghanaian policy makers expected national and corporate benefits to flow from the small but persistent relationship of military education and training exchange reveals more about their own ideologies regarding transnational educational experiences than about any mutual desire to create a "strong" military. Leaders in both Ghana's and the United States' governments willingly ignored a great deal about the other, including a reputation for neocolonial meddling, rumors of CIA involvement in Ghana's coups, and support for white rule in South Africa on one hand, and a heritage of ineffective, politically unreliable rulers, pan-African socialism, and a "malcontent" military that refused to exit national politics on the other. What both sides ignored about the other on the policy level reveals in much higher relief the faith in the fundamental goodness of the human exchange that underwrote the relationship for so long, even if both sides rarely admitted it.

Ghana's prominence as the first newly independent African state south of the Sahara made American foreign policy there ideologically representative. Americans readily took Ghana to represent the rest of the Global South, but the United States' relationship with Ghana differed from other influential or contested states in Asia, Latin America, and even Africa.[4] Even white American political leaders clamored to make Ghana a symbol of (or proxy for) their civil rights agenda.[5] Ghana also tested the integrity of the ideologies behind foreign aid that united economic development projects with security assistance but that resolutely rejected escalating weapons sales in favor of entirely human resource–centered modernization projects. Finally, American responses to recurring military coups d'état, punctuated by interludes of constitutional government in Ghana, reveal how American foreign policy evolved to tolerate military rule for different reasons over time. Because the United States' military presence in Ghana was so small and depended almost entirely on international education and training exchanges, Ghana's case demonstrates especially well the persistence and limits of ideology in foreign policy.[6] Specifically, official references to the personal interaction between Ghanaians and Americans, especially between military personnel and American communities, remained the essential justification for the American policy for thirty years.[7]

In the early 1960s, nonalignment provided Ghana's Prime Minister Kwame Nkrumah with leverage over the terms of Ghana's relationship with the United States that few other leaders of small states shared and that heightened the global propaganda value of all US-Ghanaian cooperation.[8] Nkrumah's aggressive nonaligned rhetoric irritated President Kennedy, but Nkrumah forced American leaders to recognize his power to restrict American aid that smacked of "neo-colonialism" and demand American support for projects that promised "development."[9] Kennedy made "engaging" nonaligned states the centerpiece of his new, ideological approach to foreign relations.[10] For modernization theorists in his administration, nonaligned states offered the highest psychological return on investments in the development of the "Southern Half of the Globe."[11] Meanwhile, Kennedy aggressively wooed Nkrumah through an unprecedented personal diplomacy initiative.[12] Therefore, presidents Kennedy and Johnson gauged their Cold War influence across Africa and Asia on Ghana.[13]

Later in the 1960s, Ghana manifested many of the same political and economic tragedies that Cold War competition wrought upon contested states.[14] The nonaligned coalition also fractured as Cold War tension exacerbated differences between nonaligned states.[15] Rather than focus on Ghana as a victim of Cold War competition, however, this work argues that the nonaligned movement's response to the Cold War amplified Ghana's international significance, and the United States became even more attentive to the psychological impact of its foreign policy initiatives there.[16]

In the early 1960s, American policy makers still saw the world as a national security problem, but correlated the broadly diverse aspects of the US-Ghana relationship as essentially psychological, not political, matters. US funding for the massive Volta River Dam project, Ghana's pioneering acceptance of Peace Corps volunteers, competition for leadership among UN peacekeeping forces in the Congo, and Ghana's pursuit of military training from British, Canadian, and Soviet sources all merged in a new sensitivity to the connections between anticolonial nationalism and economic development that sprung, at their heart, from challenges related to new interpretations of modernity in American foreign policy.[17] That is why American military assistance to Ghana took the form it did—limited almost exclusively to providing grants for Ghanaian officers and noncommissioned officers to travel to the United States for professional education and training. This assistance agenda persisted through the 1970s and 1980s, though, suggesting that the ideological assumptions that first connected military education to modernity had deeper and sturdier roots than other American projects abroad.[18]

European and American definitions of "modernity" had long justified imperial behavior, but the modernization that American foreign policy makers described in the 1960s threaded so-called realist concerns with national power with a new faith in the psychological transformation to modernity that social scientists proposed.[19] After 1945, American foreign policy equated the modernizing mission in contested new states into

**TABLE 1: US-SPONSORED MILITARY EDUCATION AND TRAINING, 1954–1992**

| | Number of students trained | | | | |
|---|---|---|---|---|---|
| **Region** | **1954–1962** | **1963–1972** | **1973–1982** | **1983–1992** | **Totals (1954–1992)** |
| **Latin America** | 20,017 | 43,087 (58%)* | 27,218 | 25,137 | 115,459 |
| **East Asia and Pacific (includes Vietnam)** | 77,169 | 147,990 (53%) | 51,129 | 11,107 | 287,395 |
| **Near East and South Asia** | 9,565 | 13,534 (16%) | 6,452 | 9,687 | 39,238 |
| **Europe and Canada** | 40,046 | 18,782 (17%) | 5,916 | 6,259 | 71,003 |
| **Africa** | 1,544 | 3,641 (22%) | 2,560 | 5,360 | 13,105 |
| **Ghana** | 3 | 98 (0%) | 344 | 228 | 673 |

Sources: Amos et al., *US Training of Foreign Military Personnel, Volume 2*, Appx. F; DSAA, *Congressional Presentation, Security Assistance Programs*, FY 1981, 509; DSAA, *Foreign Military Sales, Foreign Military Construction Sales and Military Assistance Facts as of September 30, 1985* (Data Management Division, Comptroller, DSAA, 1985), 84–89; DSAA, *Foreign Military Sales, Foreign Military Construction Sales and Military Assistance Facts as of September 30, 1993* (FMS Control & Reports Division, Comptroller, DSAA, 1993), 102–9.

* Number in parentheses is the percentage of students trained outside the United States before 1973. After 1975, exact numbers of students trained outside vs. inside the United States were not reported, although the majority of training occurred in the United States. DSAA, *Military Assistance and Foreign Military Sales Facts, May 1973* (Data Management Division, Comptroller, DSAA, 1973), 13.

a life-or-death struggle against communism.[20] Increasingly throughout the 1950s, a coalition of federally funded economists, sociologists, historians, and political scientists mingled around influential universities, including Princeton and the University of Chicago, and the Center for International Studies at Massachusetts Institute of Technology, to articulate new applications in "developing states" for their theory of modernization.[21] Energetic Cold Warriors and self-assured social scientists formally united in the Kennedy administration, and modernization theory guided American foreign policy with the force of ideology for nearly a decade.[22] It was through the lens of modernization ideology that Kennedy and his cabinet devised their strategy for the cascade of African states emerging from colonialism—nineteen between 1960 and 1961 alone.[23]

Despite massive upheaval in its domestic political situation and international influence, Ghana selected military personnel for US-sponsored education and training programs without interruption after 1962. This relationship was unusually persistent and constant, but it was also small; rarely did more than forty Ghanaian military personnel come to the United States in any year. Ghanaian political and military leaders claimed varying levels of autonomy in describing this partnership with the United States, but they defended their relationship and appealed to it to access other American resources. They did so because they shared a common faith that transnational education and military professionalism reinforced one another. Ghanaian and American leaders' justifications differed and evolved, especially as Ghana slowly developed its capacity to provide similar training to its own soldiers and to soldiers from elsewhere in Africa, but Ghana offers a unique view of the program's ideological assumptions that changed surprisingly little.

## Identities: Modernity, Internationalism, and Diaspora

This work posits that the best way to describe the policies and people in international military settings is by examining what they meant by modernity. Clearly, then, the terms modernity, modernization, ideology, nationalism, and internationalism require clarification.

Dense thickets of historiography obscure the term "modernity." Historian Frederick Cooper recognizes at least three conflicting historiographical senses: Modernity alternately represented Europe's civilizational legacy to the world (delivered via empire), or the bundle of Western-derived social, political, and ideological imperial constructs that "sterilizes the rich diversity of humanity."[24] Non-Western peoples might also describe modernity as a hybrid of their own cultural forms that merged their traditions with external perspectives on progress to represent their own self-consciously distinct programs.[25] This book recognizes an essential distinction between modernity (or modernization) as an ideological imposition upon African objects and modernity as a vocabulary that Ghanaians adopted in order for their claims to be recognized.[26] The same word imputes different levels of agency onto actors, but in most cases in the twentieth century, becoming "modern" equated to "improvement."[27]

Even when the actors themselves used the term, the meaning did not always speak for itself. In 1959, for example, Ghana Military Academy allotted more time to "Modern Subjects" than history, mathematics, science, or French in its curriculum.[28] In 1961, the American defense attaché to Ghana admired a textbook on "Modern Civilization" among the batch of books he presented as a gift to Ghana Military Academy.[29] In those cases, "modern" signified "legitimate knowledge," and at least three different parties in postcolonial Ghana vied for the authority to define it—postcolonial nationalists under President Kwame Nkrumah, Ghana's British-educated military elite, and American foreign policy makers.[30] In another case, when a Ghanaian army officer declared in 1969, "modern military leaders should have the breadth of outlook, cultural background and sound judgment which surpasses fundamental soldiery qualities," he appropriated the well-worn grooves of the "modern military" from American and British social science-turned-foreign policy to prepare the ground for a return to military rule, should Ghana's second republic fail to match the military's imagined aptitude for governance.[31] In the United States in the late 1950s, intellectuals behind modernization theory erected an entire taxonomy for what constituted "traditional" versus "modern"—the latter condition indicated by devotion to compound interest, Newtonian physics, and "high mass consumption"—and proposed a set of universally applicable steps to regulate the process.[32] Despite defining supposedly objective benchmarks for measuring and comparing societies in the stages after "takeoff," when modernization theorists said "modern," they *meant* "Western."[33] And when American political scientists and military assistance planners in the 1970s and 1980s routinely acknowledged no clear evidence that new states' militaries were also "modernizers," they employed the self-evident category of "modern" military technology—which only a small number of states were capable of producing—in the much less clear metaphor of what comprised "modern" political stability.[34]

"Modernization" also provides a shortcut into the historiography of American intervention in the Global South throughout the twentieth century but demonstrates how neat terms and bounded historiographies get messy when explored from the bottom up. American scholars argue that the transformation American foreign policy makers intended for the world was primarily psychological.[35] Paradoxically, then, the programs they devised to bring about this transformation were not pragmatic, as they imagined, but also reflected their own psychological assumptions about modernity and progress. That is one reason they so often used the term "developing" to describe target or client states: in order to demonstrate the process of modernization, they first infantilized those societies rhetorically. Prominent American social scientists in the late 1950s and 1960s thought they had devised a recipe for universal, transferrable stages for modernizing the "developing world," but even they acknowledged that the first acts of modernization must be personal and mental.[36] Global events in the 1960s dissolved their certainty and eroded their influence. Meanwhile, three decades of historians have questioned whether "modernization" offered anything at all to those states, or really said more about the people making the policies.[37]

Modernity also has a meaning for historians looking backward, and I use the word to describe processes of change that actually shaped people's lives even when they did not use it. C. A. Bayly argues that part of being modern was *deciding* to be modern, or at least that one lived in a modern world.[38] Historians across genres would agree on some characteristics of "modernity"—mass citizenship, industrialization, bureaucracy, and secularization within bounded nation-states in the abstract; health facilities, education, and access to markets more specifically.[39] But more than simply exposing parts of the world the West "left behind," modernity provided a language to base claims against the West—as in, "if you think we should be modern, help us find the means."[40] Ghanaian families that participated in military-sponsored international travel absorbed external standards for the modern military and the modern family, but they also employed that travel to obtain the tools to attain that modernity. This is the sense in which I employ the concept of "modernity," then: concentrating not on the theory of difference, but on modernity as a policy.[41]

Ghanaian soldiers and their families also moved fluidly between local, national, and global identities. The weak Ghanaian state's drastic fluctuations in the three decades after independence compelled such fluidity. Even the very existence of a Ghanaian "nation" during those years is far from settled.[42] On one hand, as the product of British education, training, and organization, the Ghanaian army represented both the "colonial-school-generated nationalism" and the successor/heir to the colonial system Benedict Anderson identifies.[43] Other traditions of Ghanaian nationalism, especially those Kwame Nkrumah championed, were "famously pan-African," and "rather oddly unconnected with the Ghanaian state."[44] Given such weakness in the Ghanaian state, it is not surprising to find alternate sources of military corporate identity and find them in the same place we would look for other Ghanaians: abroad.[45] Internationality, in other words, was an alternative to identities of nationalism and urbanization competing within Ghana during those decades, and Ghanaian military families expressed it in ways similar to other Ghanaians in the post-independence diaspora.[46]

Using a diaspora approach to explain how Ghanaian families crafted transnational identities from the meaning they derived from international military travel leads back to the essence of transnationalism, identity, and internationalism. Transnational histories consider the nation-state an essential layer of foreign relations and foreign policy, but not the only one, and often not the dynamic one.[47] Despite their presence in an arguably proto-national organization, Ghanaian military families abroad and Ghanaian peacekeepers regularly ascribed transnational meanings to their international experiences. They shared with other families in American communities around military schools a devotion to cultural exchange, peaceful cooperation, education, and community service. On peacekeeping deployments or on the home front during their family members' deployments, they expressed national pride sometimes but tempered it with devotion to the United Nations as the most promising agency to deliver world peace. By some definitions, such alternate, transnational routes toward peaceful cooperation are part of the definition of internationalism, although few scholars recognize international military partnerships as legitimate participants.[48] The actors in this work never wholly escaped the shadow that the United States, Ghana, or the United Nations cast over their options, and so it is still appropriate to call this an international, rather than transnational, history.[49]

Internationalism generated via military travel suggests another avenue for examining organizations that reflect transnational concerns and strengthen the sense of global, human interdependence. Akira Iriye proposes, and I agree, that what others think were the geopolitical triumphs of the "real world" in the twentieth century were actually short detours on the way toward a global "imagined world."[50] In this light, the temporary dominance of modernization theory over American foreign assistance policy in the early 1960s, military coups in Ghana in 1966, 1972, 1979 and 1981, and the military balance of power between Israel, Egypt, and Lebanon may not have been the "real world" at all, but diversions on the road to the imagined world Iriye conjures. That helps explain why measuring and justifying the United States' international military education program so befuddled policy makers for decades.

Such a new understanding of the "real world" sets us looking for different evidence and provides us with a different standard to measure "results" in American military assistance policy in the 1960s, 1970s, and 1980s, Ghanaian military participation in training, education, and peacekeeping abroad, and American communities embracing international military families. Senators and the State Department alike found little evidence that American international military education and training programs brought "political development" to target states in the Global South, created "professional" or nonpolitical militaries, or even bought Cold War allegiance.[51] Neither do I. Education abroad or peacekeeping service alleviated some Ghanaian soldiers' and their families' economic and political stress in that time, but not permanently, and it did not restore democratic rule promptly. Instead, I see evidence of the imagined world appearing, or reappearing, as military travel connected individuals and groups from different lands who developed alternative communities based on their cultural interchanges.[52]

Nation-states had only partial control over the content of those exchanges. Even when there were formal, state-driven curricula of knowledge to be imparted, the state, the community, and the individual families themselves interpreted the context of the exchange for themselves. Indeed, the state seemed to acknowledge and, to some extent, celebrate its inability to regulate this context, reflecting an internationalist faith. This paradox—states' ultimate lack of authority over the psychological essence of international military travel—is one of many that helps us better understand military members' internationalism.

As a social history of military subjects, some of the military terms below are intentionally imprecise. Staff college, for example, meant different things in different countries. In the United States, it generally applied to schools for midgrade officers, usually majors and lieutenant colonels (i.e., Command and General Staff College, Air Command and Staff College, Marine Corps Command and Staff College). These were often divisions of larger military educational structures, such as the Naval War College, Air University, or junior partners to other schools, such as the Army War College. The situation was similar in Ghana. The Ghana Armed Forces Command and Staff College included both a junior division and a senior division. The junior division began in 1963; the senior division (for majors and lieutenant colonels) received its first students in 1977. For the purpose of this book, in order to keep the terms roughly similar, I use the term "staff college" to refer to the senior division in Ghana and the intermediate service schools like Fort Leavenworth's Command and General Staff College in the United States or their equivalent in allied countries.

Similarly, the "army" and "armed forces" were not the same either in the United States or in Ghana, but I sometimes interchange the terms here. Ghana has had a separate air force and navy since 1961, but they have always been small, roughly 10 percent of the army's size, which varied between about 7,000 in the early 1960s, grew to 14,600 by 1966, then slowly shrunk to 10,000 by 1981.[53] All three shared Ghana's military academy and staff college, although all had separate training facilities as well. I do not intend to gloss over institutional differences between the services, especially since these differences are even more significant in the very large armed forces of the United States. This book explores international military education from a national policy perspective and the communities around certain military facilities, however, not distinctions between the different services.[54]

Finally, written sources about Ghanaian military subjects are scarce and problematic. I received research funding from the United States Air Force Academy's directorate of education and coordinated access to Ghana's staff college, director of public relations, and military academy through the US Embassy's defense attaché office. I recognize that these military-sponsored resources demand careful attention to objectivity.[55] But the sources' own patchiness do as well. Even within the Ghana Armed Forces' collections, access was gravely limited to extant collections (which were few), to the willingness of Ghana Armed Forces personnel to cooperate (which they seemed exceptionally willing to do), and the likelihood that little evidence that openly criticized Ghana's military regimes would have reached me, if it survived at all. The directorate of public relations for the Ghana Armed Forces maintains only a fraction of the *Ghana Armed Forces News* editions they published between 1967 and 1992. Other libraries around the world maintain other issues, but not with regularity. Although Ghana published journals from all of their peacekeeping contingents after 1979, only a few before 1990 survive. Finally, in all these official publications, women rarely show up, and only filtered versions of their words appear printed in official armed forces publications. Memoirs (by male authors) describe Ghanaian families' economic and political strategies *as families*, but Ghanaian military wives' voices or Ghanaian women soldiers' voices from this time period are very difficult to locate. On the other hand, if Ghanaians' military service abroad resembled other Ghanaian migrant or traveling workers in other contexts, we recognize common strategies for Ghanaian military wives and Ghanaian women elsewhere in the diaspora.[56]

There was no necessary connection between what the Ghanaian officer experienced and what the United States expected its international officers' program to achieve as an application of foreign policy. After 1966, Ghanaians as a whole may have quietly questioned the military's legitimacy despite (or because of) its near-continuous hold on Ghana's government, but the United States never discontinued its programs of military educational assistance. Abroad, Ghanaian officers' status as students and recipients of international education did not change. Neither did their "mission" to translate those lessons back to Ghana, regardless of the uncertain legitimacy of military government. This constancy reveals as much about what the United States considered the role of the military in the modern nation-state as it does about Ghanaians themselves or the Ghanaian army. How Ghanaian officers interpreted the experience or how it provided individual, corporate, and national resources in the form of a transnational or global identity, is a different question, wrapped up in the meanings of nationalism, modernity, and identity itself.

# 1 Modernization Ideology and the United States Military Assistance Program

*Military leaders are often far less suspicious of the West than civilian leaders because they themselves are more emotionally secure. This sense of security makes it possible for army leaders to look more realistically at their countries. All of these considerations make it easier for military leaders to accept the fact that their countries are weak and the West is strong without becoming emotionally disturbed or hostile toward the West.*

– Lucian W. Pye, "Armies in the Process of Political Modernization"

"With the signing into law of this bill, a Decade of Development begins," President John F. Kennedy declared on September 4, 1961, as he authorized the Foreign Assistance Act of 1961.[1] Seven months had passed since he took office, and in some ways Kennedy's "Decade of Development" was already in full swing. Within weeks of the inauguration, his "Special Message to the Congress on Foreign Aid" and urgent-sounding State of the Union address had replaced "mutual security" with "development" as the justification for American foreign assistance, had claimed "the whole southern half of the globe" as the new battleground in the Cold War, and had offered a compelling litany of priorities for the fight.[2] Atop the list, Kennedy blasted the United States' $3.2 billion security assistance program as "bureaucratically fragmented," "awkward and slow," "haphazard and irrational," "obsolete, inconsistent and unduly rigid," and he recommended "the replacement of those agencies with a new one—a fresh start under new leadership."[3] The Foreign Assistance Act was an essential first step, and all summer, Secretary of State Dean Rusk, Secretary of Defense Robert McNamara, and assorted undersecretaries and deputies wrangled with congressional committees to ensure its passage.

Kennedy issued three orders the following day that demonstrate how tightly he correlated the Cold War to military assistance for the "southern half of the globe." After ordering the resumption of underground nuclear testing, Kennedy demanded an update on the status of education and training programs for military officers from Latin America and Africa.[4] Responding to a political crisis in Brazil, Kennedy announced, "As the events of the past week have shown in Brazil, the military occupy an extremely important strategic position in Latin America. I would like to know how many officers we are bringing up from Latin America to train here and whether we could increase the number."[5] He issued a national security action memorandum requesting "what other steps we are taking to increase the intimacy between our Armed Forces and the military of Latin America" and recommended creating programs to bring to the United States "a good many officers from the different countries of Latin America" for lessons in combating communism and subversion, controlling mobs, and fighting guerillas.[6] "In addition to increasing their effectiveness," Kennedy supposed, "it would also strengthen their ties with the United States."[7] As if completing that thought, Kennedy issued National Security Action Memorandum 89 moments later, demanding an update on "whether Mr. Nkrumah is going ahead with his plan to send 400 of his troops to the Soviet Union for training."[8]

Those two concerns—Latin American officers training in the United States and the source of military training for Ghanaian cadets—reflected both old and new directions in US foreign policy in 1961. Providing training for military personnel in target countries had been an essential aspect of American military assistance in the decade prior to the "Decade of Development," but Kennedy's idea to bring as many of them as possible to the United States reflected a new faith in the potential for the United States to instill more than just combat skills. Likewise, Kennedy and his cabinet considered Ghana a bellwether for all of Africa. They acknowledged President Kwame Nkrumah's influence as a pan-African leader, and carefully considered the public relations impact of every aspect of US-Ghanaian relations.[9] Kennedy and Nkrumah had tangoed all summer. In March, after meeting Kennedy in Washington, Nkrumah declared, "We are anti-colonialists and we shall always remain so until all the colonialists are gone" but, to calm US fears, added, "I think anti-colonialism was invented by the United States."[10] In April, Nkrumah provided Kennedy a "small miracle" by becoming the first world leader to invite Peace Corps volunteers, probably saving Kennedy's high-profile program.[11] By September, Nkrumah pressed Kennedy to confirm the United States' offer to help fund the Volta River Dam project while simultaneously offering (or threatening) to send up to four hundred army cadets to the Soviet Union for training.[12] All sides accepted that military training abroad denoted states' strategic alignment in the Cold War, but, more than that, Kennedy and his advisors believed that military training caused lasting psychological and emotional transformations that could propagate throughout a society.[13] This was why they were so keen to maximize the number of officers from Latin America attending training in the United States and so concerned over Nkrumah's plan to send cadets to Russia.

This chapter explores how training and education for foreign military personnel in the United States came to assume such strategic and psychological significance in the early 1960s. It argues that influential academic elites, guided by modernization theory, integrated prior American assumptions about the military as a modern institution in the late 1950s, and they restructured foreign assistance policy in the 1960s to make international military education and training a preferred method to transform visiting elites from target states into "modern" men and American allies at a low cost. These programs tapped into the deeply held assumption that life among American communities spontaneously conveyed the same political, economic, and social values that American social scientists later equated with modernity, and that helps explain the persistence of international military education programs long after modernization theory declined in influence.

American military assistance programs had implicitly accepted American communities' power to transform visiting military personnel from underdeveloped countries for half a century. That faith paralleled other American assumptions about the psychological power of military defeat to goad the underdeveloped world to choose to modernize.[14] These two beliefs merged powerfully in the late 1950s, as American military assistance programs expanded rapidly in the Cold War's heat and as a coalition of professional academics and policy makers applied new theories of economic development to American foreign policy. With funding from an often-opaque alliance of philanthropic foundations, research universities, and the CIA, a core of influential historians, sociologists, economists, and political scientists developed a theory—or ideology—of modernization, which articulated a collection of assumptions about the nature of American society and its ability to transform the economically and culturally deficient world.[15]

The great, founding projects of modernization theory connected economic development in poorer nations to eventual success in the Cold War, but they were not, essentially, about economics.[16] Modernization was both an ideology and a discourse, and it dictated strategies in foreign aid, trade, counterinsurgency, and nationalism that were partly economic, partly political, but largely mental. Academics such as Walt Whitman Rostow, Lucian Pye, and Daniel Lerner became policy makers after Kennedy's election in 1960, and they applied modernization theory to all aspects of American foreign policy as a result.

When modernization theorists in the Kennedy administration asserted control over military assistance policy, especially those parts of American military assistance that sponsored foreign military personnel for training in the United States, they were not inventing a new purpose for military assistance. Instead, they reimagined the role for military forces in newly decolonized states away from intimidation and coercion into a kind of institutional bridge. A new ideological emphasis on economic and political development could supplement preexisting programs that connected the US Armed Forces with the new states' national militaries. Development theorists used American military assistance programs as vehicles to transfer "modern" ideas about technology, rational organization of society, and citizenship directly from westerners onto new states' emerging military elites.[17] In the larger context of the 1960s, which often rationalized military force in the service of modernization projects, programs that targeted specific military individuals for lessons in modernity thrived.[18]

This chapter describes the two essential transformations that had to occur in order for this to happen. First, modernization theorists wrested control of the massive American Military Assistance Program of the 1950s from "unimaginative" and security-minded bureaucrats in the Eisenhower administration and Defense and State Departments. Second, those theorists supplied military assistance with a new reserve of intellectual credibility—or at least the opportunity to reframe the American Cold War foreign policy debate away from a defensive focus on "security" to match their program of transforming the world through economic development.

Modernization theory fueled the great, often tragic projects of American foreign policy in the first half of the 1960s, but behind the projects lay assumptions about what made modernized *people* and what role those people played in hoisting their societies into modernity. Massive development programs, including the Alliance for Progress, TVA-style resettlement in Afghanistan's Helmand Valley, and the "strategic hamlets" on the Mekong River, typified the lofty, often faulty, ideologically driven "Economists with Guns" at the tragic height of their power.[19] In the heyday of authoritarian high modernization, there was no distinction between hydraulic and social engineering, dam builders achieved the status of a technocratic warrior elite, ideology was strategy, and "developmental dictatorships" were preferable to vulnerable democracies.[20]

Not all projects of modernization theory occurred overseas or with such tragic consequences. American military assistance, including education and training of foreign military personnel, occurred in hundreds of places. For many soldiers, especially in Southeast Asia and Latin America, American military training came to them.[21] For thousands of others, officers and enlisted—up to fifteen thousand per year, some years—the United States paid for their travel and training in America. Especially in countries where the United States had few military assets, little capacity to project military power, but the desire to influence policy, international military education and training was the only program of American military assistance. The United States had such a relationship with Ghana, Nigeria, Mali, Senegal, Niger, Ivory Coast, and Liberia between the early 1960s and 1990.[22] In Ghana, despite twenty-five years of military regimes—some with American support, some without—alternating with brief interludes of civilian government, American military assistance policy based solely on training exchanges sturdily weathered great changes in American foreign policy objectives and Ghanaian political realities.

The Vietnam War supposedly exposed the hubris behind modernization ideology. Massive aid budgets collapsed and the "New Mandarins" retreated, chastened, to academia.[23] But one of the most compelling ideas behind the ideology in the first place never went away—that life among ordinary Americans left lasting impressions on visitors from the Global South, and these individuals could, over time, refine entire societies.[24] That vision of the "imagined world" of global, human interdependence the modernization theorists had temporarily assimilated as they tended the supposedly "real world" of competing sovereign states, power, and nations, had survived. It was not clear whether that ideal would be enough to preserve international military training programs in the era of reduced faith in foreign aid, reduced confidence in the good intentions of the US government, and reduced budgets after the Vietnam War. Individuals around American schools always considered their relationships with visiting military families the program's most significant outcome anyway, so they seemed not to notice the rise and fall of the sociology of modernization theory.[25]

## Mutual Security Program Reform as Ideological Reform

With large budgets and little data demonstrating their effectiveness, American military assistance programs were likely targets for modernization theorists' intellectual makeover in the 1960s. Military assistance had been the focus of postwar American aid in Europe, Iran, Korea, China, and the Philippines since World War II, and, by the 1950s, promised to expand everywhere else the Truman Doctrine grasped.[26] In the 1950s, states in Asia,

Africa, and Latin America that were emerging from colonialism or colonial-type exploitation were central to US foreign policy. Presidents Truman and Eisenhower deemed controlling access to its resources, markets, and labor crucial to containing the Soviet Union and defending US hegemony everywhere.[27] Therefore, they justified most foreign aid primarily in terms of political and economic security, even when the targeted recipients had no obvious military connection.[28] Nearly all foreign aid deemed even remotely "security-related" fell under the massive Mutual Security Program, which had grown to $3.2 billion by 1960.[29] Mutual Security Program projects designed to increase international cooperation and foster peace dwarfed any comparable civilian program. In 1960, for example, the Mutual Security Program provided funding for 16,500 foreign military personnel to receive training or education in the United States—exceeding the number of students trained there under the Fulbright, Smith-Mundt, and Agency for International Development programs combined.[30]

In the first months of Kennedy's presidency, when he promised to transform the United States' bloated foreign assistance bureaucracy and replace it with a responsive, pragmatic, and administratively centralized infrastructure headed by "first-class development planners," the Mutual Security Program was a prime target.[31] As he eliminated the jumble of disparate assistance programs it unevenly contained, and consolidated control over all foreign assistance—civilian and military—within the State Department's new Agency for International Development, his staff reevaluated the ideological potential of American military assistance and altered those aspects they considered most likely to have long-term effects at the lowest cost. Programs that sponsored foreign military personnel for training and education in the United States survived because they could provide both.[32]

Kennedy was not the first president to think seriously about foreign aid, however, and while his reforms were significant, they occurred at the nexus of contending political forces over a decade old. During his second term, Truman attempted to replicate the successes of the Marshall Plan, proposing a massive foreign aid program to combat communism by promoting economic recovery through "technical assistance" to poor nations around the world, but a parsimonious Congress immediately limited the program's funding.[33] Eisenhower likewise struggled to convince Congress and the American people of the need for more foreign aid funds and greater flexibility to administer them. Different congressional factions hindered Eisenhower's initiatives to expand the foreign aid program while pressing the administration to spend its money on programs variously more, or less, justified by national security alone—even as the very definition of "national security" shifted and expanded continuously.[34] On the other hand, nearly all agreed that the American people themselves were the best ambassadors of the American way of life to the rest of the world.[35] Notable cultural diplomacy programs expanded in the 1950s, including the Fulbright-Hays scholarships; the president's Special International Program—jazz musicians on tour; and "People-to-People International," which Eisenhower founded in 1956 to help "people to get together and leap governments—if necessary to evade governments—to work out not one method but thousands of methods by which people can gradually learn a little bit more of each other."[36] And since any dollars that brought foreigners to the United States would be spent in the United States, each party in the negotiation over the size and role of American foreign assistance policy considered those dollars well spent.

President Eisenhower faced withering congressional attacks as he attempted to reform the Mutual Security Program during his second term.[37] Not all angles played equally effectively in public opinion, but with $3.2 billion appropriated for a Mutual Security Program that encompassed military assistance, development loans, technical assistance ("Point Four"), and the nebulous category "defense support," the charge that the United States' foreign aid programs were too expensive consistently resonated with the American people.[38]

Criticism generally came from two opposite camps: those who objected to foreign aid in principle and those who thought America's foreign assistance program needed to focus less on security and more on development. Eisenhower believed that the American people fully accepted the responsibility for military assistance in order to contain communism, but he thought they did not comprehend how broad the definition of military assistance could be.[39] He was unable to defend the nation's interests and contain communism abroad, he complained, while congressional expectations for budgeting and oversight rose and American public opinion strongly opposed foreign "economic aid."[40] He dreaded a retreat from America's proactive financial role abroad, telling Congress "the only alternative we can see to the inter-dependent allied Free World, strengthened by our aid where needed, would be the Fortress America concept—taking our first stand in the last ditch."[41]

American public opinion had never fully embraced foreign assistance, as many in Congress routinely remembered. A trail of committee hearings and reports criticized the Mutual Security Program for a host of reasons: it wasted money on poorly conceived or elaborate programs, it funded plans that produced no tangible results, American loans would likely never be repaid, it spent dollars that provided little help to the United States' troubled economy, and it played into Soviet rhetoric of American militarist expansion. Senator

Allen J. Ellender (R-LA) colorfully observed, "A decade of many ill-conceived and loosely managed expenditures of America's wealth" had turned "what began as a nourishing broth of 'international cooperation for peace' [into] a witches' brew—a serious threat to our own economic stability and growth."[42] William Dawson's (D-IL) House Committee on Government Operations, attacked the entire aid agenda. "It is, quite literally, all things to all men," he thundered in 1958, claiming he had "never seen *any* evidence that the level of aid for *any* country [had] been systematically determined by economic experts."[43] Representatives from all sides contended that the executive branch had too much authority to act without congressional consent.[44]

Others in Congress did not shrink from the idea of massive aid but objected to the program's execution, which seemed to lack a central strategy. A powerful new group of senators—including John F. Kennedy—pushed for even more proactive foreign assistance, deemphasizing military assistance and focusing more aid specifically for economic development.[45] "Foreign aid is both an unavoidable responsibility and a central instrument of our foreign policy," J. William Fulbright's (D-AR) Senate Committee on Foreign Relations acknowledged in 1961. "It is dictated by the hard logic of the cold war and by a moral responsibility resulting from poverty, hunger, disease, ignorance, feudalism, strife, revolution, chronic instability, and life without hope."[46] Foreign aid was "indispensable," the Senate concluded, but "it can also be said that it has fallen short of the millions of words that have been uttered in its behalf."[47]

Eisenhower's attempts to reform military assistance made little headway and only confirmed this group of progressives' misgivings. In September 1958, Secretary of State Dulles warned Eisenhower, "Unless we are able to justify a military assistance program in forthright and explainable terms . . . we will be faced with increasing pressures for indiscriminate cuts in military assistance funds in the interest of devoting the resulting savings to economic development programs."[48] Before asking Congress for another dollar for the Mutual Security Program, Dulles recommended that "the basic purposes of our military aid, and the standards to be used in fixing its level, be appraised by a public committee of respected and qualified private citizens."[49] In November 1958, Eisenhower tapped his friend William H. Draper Jr., a former army general, former undersecretary of the army, and "the first professional west-coast venture capitalist" to head a special committee to study the Military Assistance Program.[50] Following his dream of shifting the country's financial core west, Draper formed the venture capital firm Draper, Gaither and Anderson in 1958, joining retired air force major general and former ambassador to NATO Frederick L. Anderson Jr. and Horace Rowan Gaither

Jr., chairman on the boards of the Ford Foundation and RAND and author of the influential and grim 1957 Gaither Report on nuclear survival.[51] In his army days, Draper had helped administer the Marshall Plan in Europe, and his conclusions reflected this experience.

The Draper Committee was another of the ad hoc committees of former generals, industrialists, and businessmen, that had become a kind of shadow branch of the US government after 1945 and which drew the disdain of intellectuals like Rostow.[52] Eisenhower, he later accused, was "shaped by military and budgetary tactics rather than a theory of history."[53] He was "unimaginative," conservative, willing to "innovate as little as his responsibilities required," and motivated only to "reduce the burden in cost and manpower of holding the line around the periphery of the communist bloc."[54] To Rostow, Eisenhower relied on "mediocre technical assistance types" like Draper, with military and business backgrounds, but was "desperately short of men at home and in the field who understand the economic development problem."[55] Under Kennedy, prominent economists, sociologists, and historians from MIT's Council on International Studies re-formed the powerful cliques that wielded similar influence over foreign assistance policy.

Embedded in their final report, published in August 1959, the Draper Committee revealed a special enthusiasm for training foreign military personnel in the United States. They also recommended the expected things: consolidate aid programs, strengthen the State Department's authority over nonmilitary assistance, extend aid appropriations for two or more years instead of just one, and require the Department of Defense to fund those military assistance programs it desired out of its own budget.[56] But programs that sponsored training for foreign military personnel in the United States received Draper's most significant recommendation. In 1960, $99 million of the $3.2 billion mutual security budget went toward training foreign military personnel, he claimed, yet "there is no single aspect of the Military Assistance Program which produces more useful returns for the dollars expended than these training programs."[57] Despite Draper's strong criticism that recommended abandoning other aspects of the Mutual Security Program, these training programs gathered increasing praise.

Even Draper admitted, however, that in the hands of military trainers, these programs failed to fulfill their potential.[58] Military training needed to broaden its scope to coincide with the overall objectives of American foreign policy and focus on building leadership and professional expertise, and not simply transmit technical skills.[59] "In some of the less developed countries," Draper claimed, "a large proportion of the managerial and executive skills that exist are in their armed forces."[60] More than a military training program, therefore, the Military Assistance Program needed to embrace its potential for sociological transformation. Through it, "recipient countries undergo significant changes—organizational, sociological, economic, and sometimes political."[61] The committee noted that the program brought more than ten thousand foreign nationals to the United States each year, but recommended that it expand to include more. "This training," Draper argued, "can do far more than teach recipients to use military equipment and research. It brings foreign nationals into close contact with United States citizens under conditions which tend to promote an appreciation of the values of our civilian way of life."[62] Policy makers in the Kennedy administration later seized this sentiment and codified it into policy.

Draper still did not stray far from inherited, strategic justifications for sponsoring foreign officers in the United States. "Many officers, returning to their native countries, often form the nuclei of Western-oriented leadership," his final report argued, as if being Western-oriented sufficed as the goal in the Cold War.[63] Draper's outlook was primarily security oriented, but he recognized the potential for military training exchanges to develop individuals' understanding and respect for American institutions and acknowledged those individuals' potential to transform their native military, social, and political institutions along more modern lines. This potential, it turns out, was precisely the point upon which the "old look" of the Eisenhower administration would pivot to the new approach favored by Kennedy and his cabinet of academics–turned–policy makers.[64]

## The Modernization of Military Modernization

In August 1959, in the same month that the Draper Report spelled out the Eisenhower administration's final attempt to reform the massive Mutual Security Program, the RAND Corporation sponsored a conference in Santa Monica, California, on the role of the military in "underdeveloped" countries. Some of the most influential academic proponents of modernization attended, including Lucian Pye, who was raised by Protestant missionary parents in China and whose history of the Malayan insurgency in 1956 prescribed the psychological process of modernization as the solution for the economic and social appeal of communism; Edward Shils, a prolific scholar of Cold War foreign policy, propaganda, and new nations; and Daniel Lerner, whose 1958 study of Turkey, Egypt, Jordan, Syria, Iran, and Lebanon in *The Passing of Traditional Society* first explicitly theorized the process of social transformation as "modernization."[65] All three worked for MIT's Center for International Studies (CIS) at the time, the CIA-sponsored research center that was also home to the influential economist Max Millikan and to Walt W. Rostow, who eventually rose to chair the State Department's policy planning staff under Kennedy and who served as national security advisor under Johnson.[66] The conference proceedings, published in 1962 as *The Role of the Military in Underdeveloped Countries*, became the most influential attempt to come to grips with the paradox that transitional societies "apparently find it easier to create modern armies than most other modern structures."[67] At that moment, social scientists informed by modernization theory seized the initiative in theorizing, rationalizing, and executing American military assistance. As they did, programs that brought military officers from new states to the United States for education and training acquired special significance because these programs encompassed so many of modernization theory's tenets of faith—at a low cost.

* * *

The historiography of modernization and development ideology clearly establishes the preeminence of social science in driving American foreign policy toward developing states. Social science was slow to incorporate military assistance, however, which had thrived under a different set of assumptions during the second half of the century.[68] In the late 1950s, modernization theorists successfully absorbed the rationale and administrative control over military assistance for three reasons. First, congressional and American public opinion supported military assistance to the Global South but regularly disapproved of providing strictly "economic" aid, so modernization theorists carefully couched their programs in terms that permitted access to large "security" budgets. Second, even though social scientists essentially refused to examine the significance of the military in newly decolonized nations prior to 1960, many other Western scholars already acknowledged that the military often played a "modernizing" role in new states. Finally, a series of significant scholarly works in the late 1950s and early 1960s filled the gap between the undertheorized phenomena of military modernization and the spread of military rule in the Global South.

Influential academics–turned–policy makers supplied prior explanations of the military's modernizing function with intellectual respectability and a share in the political bonanza for modernization theorists in the Kennedy administration. The way international military training programs unified the psychological aspects of modernization—a person could become a modern person, and the good soldier was a modernized man—with the prior and resilient public faith in the goodness of American communities later insulated these military assistance programs when modernization theory declined.

The myth that military conquest begot modernity in Asia, Africa, Latin America, and the Middle East had a long history. Since the mid-1800s, both Western and Egyptian nationalist historians agreed that Mehmed Ali's conscripted army, formed in response to Napoleon's humiliation of Egypt, laid the foundation for Egypt's national "take-off" in the early nineteenth century and earned Ali the title "Founder of Modern Egypt."[69] Even then-colonel Anwar Sadat apparently considered the criterion of a good Egyptian officer in 1960 "the degree of his familiarity with the West."[70] In 1929, Arnold Toynbee mused before the Royal Institute of International Affairs that the people of Turkey, Egypt, and Persia owed their "present enlightened rulers to the Great War."[71] Military defeat in World War I had awakened a spirit of self-questioning, dislodged "traditional Islam . . . in its bigoted form," and "given way to a spirit of inquiry and self-questioning" in the Middle East.[72] In Turkey between the wars, but especially after World War II, secular, intellectual politicians had harnessed the reform-minded military to produce a West-facing, rapidly modernizing, civilian-run state, and an essential Cold War ally.[73] Similar events seemed to unfold in Iran after World War II.[74]

Military conquest also fueled the United States' modernizing mission abroad. A *New York Tribune* reporter sailing with Matthew Perry in 1853 exulted at the "great American Navy—that glorious institution which scatters civilization with every broadside and illuminates the dark places of the earth with the light of its rockets and bombshells."[75] Modernity seemed to pour from American cannons! Japanese annexation of Korea in 1910 earned Western commentators' praise as a laudable attempt to spread modernity.[76] After the United States occupied the Philippines, Cuba, and, later, Haiti, the military led reconstruction projects that imported modern technology.[77] The army and navy also worked closely with other aspects of the modernizing mission, including missionaries, philanthropic organizations, and, eventually, development theorists, who arrived after the dramatic displays of American power that reoccurred in Tokyo Bay, Manila, Haiti, and postwar Europe in the century before 1950.

Americans offered military education in the United States as one path for elite subjects of its expanding empire to transform their societies in the early twentieth century. While the US Army initiated many infrastructure projects in the Philippines as part of its counterinsurgency strategy, its overall program was one of "social engineering."[78] To American commanders in the Philippines after 1898, native Filipinos became fit for military leadership only after the US Congress authorized admitting Filipino cadets to the US Military Academy at West Point in 1908.[79] Filipino West Point graduates could then command indigenous scout units, liberating the US Army from its ongoing pacification mission in the Philippines. In 1935, American general Douglas MacArthur envisioned a Philippine military academy "built on the lines of West Point," and convinced President Manuel Quezon that only Filipino graduates of the real West Point and Annapolis become the first superintendent, commandant, and training staff for the new Philippine Military Academy. Between 1904 and 1937, the United States sponsored ninety-three Chinese men to attend American military colleges—including West Point, Virginia Military Institute, The Citadel, and Norwich—to help modernize nationalist China's army and insulate it from communists.[80] In 1919, acting Secretary of State Frank L. Polk pleaded with Congress to permit "intelligent and spirited young men" of the "sister Republics of the Americas" to send cadets to West Point and Annapolis, anticipating that they might "return to their native countries deeply impressed by our system of training and our institutions and well prepared to develop military training at home."[81] Congress finally acquiesced in 1938.[82] Military education and training in the United States had deep roots in American modernization programs abroad; its importance expanded in proportion to Cold War military assistance programs.

The truce that emerged in the late 1950s between prominent social scientists and modernization theorists who advocated military assistance as a strategy for development was so important because, prior to then, most thoughtful observers did not equate military capacity with modernity or progress and warned of the ascendance of military ideology over the state. From Princeton University, the influential historian and German émigré Alfred Vagts warned in 1937 (and again in 1959) against militarism—evident in the spectacles of fascism—that was antidemocratic and recent, but not modern.[83] Armies served modernity only when they made "efficient, rational, up-to-date, and to a certain extent, humane uses of the materials and forces available to them," Vagts wrote.[84] Militarism, on the other hand, was antirational, traditional, and displayed "the qualities of caste and cult, authority and belief." Even after World War II, Vagts lamented the spread of "civilian militarism," evident in "the churn of officers rising in Turkey, Iran, Thailand, Indonesia, Egypt, Iraq, [and] Lebanon and civilian governments appropriating military titles for their prominent civilians in the Soviet Union and other places."[85] Most social scientists prior to 1960 shared Vagts's reluctance to embrace the union of militarism and modernization as a promising strategy for encouraging development in poorer states, motivated either out of a traditional liberal suspicion of armies, by a presumption that military phenomena were anachronistic, or by preoccupation with other trendy social issues like industrialization and urbanization.[86]

What was different about the modernization theorists' understanding of military modernization was the evolving definition of "modernity." Since the seventeenth century, Europeans had equated their technology—especially their military technology—with social and cultural superiority.[87] By the nineteenth century, influenced by theories of evolution, the "civilizing mission," and abstract notions of equality that accompanied industrialization and urbanization, "modern" became synonymous with "improvement" while simultaneously implying "backwardness" for the "primitive societies" abroad.[88] The Americans in Tokyo Bay in 1853, Manila in 1899, or Panama in 1912 were no different.[89] Still, "modernization" before the 1950s usually referred to technical improvements in physical or bureaucratic infrastructure, administrative streamlining, or the introduction of new machine equipment or chemical fertilizers, for example.[90] To Edward Shils, modernity demanded "breaking the power of traditional interests of chiefs, sultans, and priests, universal public education, democracy, being scientific, [and] being economically advanced and progressive"—in other words, "'Modern' means being Western without depending on the West."[91]

To the extent that people in the new states aspired to be "modern" before the 1960s, they were also reacting to the spread of liberalism, nationalism, the concept of class, and world religions in an increasingly globalized political and cultural economy.[92] By negotiating these influences for themselves, people in developing places could decide that they were modern, or that they lived in a modern world, whether they liked it or not.[93] Before World War II, the term "modernization" was never used to describe a society as a whole.[94]

This is not to say that all social change caused by war was modern or represented modernity. Instead, when specific scholars of modernization theory, such as Lucian Pye, Daniel Lerner, or Edward Shils, accepted military technology and military sociology as contributors toward the economic, political, and mostly psychological evolution modernization theory encompassed, they modified prior understandings of modernity to fit a new paradigm. Modernization theory not only described, it could now prescribe how entire societies transformed.[95] In Lerner's words, "Individuals and their institutions modernize together."[96] By the late 1950s, modernization as an ideology had coalesced around deeper, older cultural assumptions about American society that intellectuals, officials, and the American public already largely shared.[97] As its theorists became more influential, they also incorporated prior assumptions about the role and promise of American military assistance programs.

The August 1959 RAND conference in Santa Monica, California, was the turning point in reinterpreting the modernizing potential for the military in "underdeveloped countries" because it separated the economic problem of military spending from the sociological potential for military forces to transform societies. Prior to that, economists condemned military spending for derailing, not helping, developing economies. Max Millikan and Walt W. Rostow's 1957 *A Proposal: Key to an Effective Foreign Policy*, the most influential work on development economics thus far and which claimed to have harnessed the near-unanimous, collective expertise of the CIS, baldly declared that military spending prevented poorer states from growing economically and only made their leaders more likely to pursue totalitarian measures.[98] In economic policy, military spending was suspect. As part of modernization theory, on the other hand, which proposed a series of steps toward economic "takeoff" that required first the social transformation from "traditional" to "modern" personalities, programs that isolated future military leaders from their "traditional" societies could be a perfect fit.[99] At this 1959 conference, social scientists blessed that essential intellectual leap over the economics of development to the uses of foreign military assistance to create modernized people.

The military in postcolonial societies could be an avenue for advancement and a protodemocratic institution, these scholars argued, but the potential for abuse of this advancement made it essential to understand its influence. Distinguishing "civilian" and "military" outcomes from foreign assistance was difficult enough. American scholars did not agree on the stakes. At the RAND conference, University of Chicago sociologists James S. Coleman and Belmont Brice Jr. argued that in African states, civilian policy makers and the army shared the same basic goals of stability, order, national unity, and rapid modernization. "These are the goals of the educated African nationalist," they wrote, and if the civilian regimes could not deliver them, "there is a high probability that the military will intervene."[100] These scholars acknowledged "a unity in the common economic and social problems" and the "paramount task of finding a format for civil-military relations appropriate to their social structure."[101] "Military aid has had to become economic aid," Shils wrote, since it already contributed to "roadbuilding, health facilities, communications networks and the like, all of which have directly facilitated economic growth."[102] Shils, Pye, Coleman, and Brice all assumed that military assistance had the potential to introduce economic and institutional reforms that yielded modernization, but only when that assistance behaved like economic aid and not as an incentive for national militaries to intervene into politics. In 1959, as African states seized their independence in great numbers, widespread military intervention was still a concern but not yet a reality. It became reality across Africa and Asia early in the 1960s, but, in 1959, the example of Turkey, where the modernizing military spurred advancement, then retreated from politics, held great promise.[103]

Participants at the 1959 conference lamented how few scholars had seriously examined the military's potential political influence on African and Asian "new states."[104] Lucian Pye's presentation, reprinted in *The Role of the Military in Underdeveloped Countries*, lowered the barriers to the scholarly acceptance of military assistance in the service of modernization.[105] Although he later taught courses in counterinsurgency for the State Department, Pye was not simply an apologist for military rule.[106] He regarded military dictatorships as interim solutions, at best.[107] He pragmatically pointed out, however, "The military has become the key decision-making element in at least eight of the Afro Asian countries."[108] He continued, "We are confronted with the awkward fact that there has been almost no scholarly research on the role of the military and the political development of the new states."[109] Daniel Lerner considered "the absence of definitive case studies" an opportunity to "make general observations" that offered considerable tolerance for early military participation in the process of modernization but that, it later showed, rarely applied outside of Turkey.[110] The scarcity of data seemed to result from a willful refusal to integrate military sociology into political science, not just an imbalance in the study of social change in the Global South.[111]

Within ten years, most observers would decry these early attempts to justify or co-opt the momentum of military assistance for advancing modernization theory as unfortunate or sinister evidence of the "theory-research gap" between expectations and political, social, and economic reality in the new states.[112] In 1959, however, political conditions in the United States made social scientists reconsider whether military elites in new states might be a fertile field for their modernization efforts.

Some scholars considered the military the ideal example of modernity in developing states because the ideal soldier supposedly manifested a complete transformation into a "modern" personality, which valued technology, industry, mobility, and nationalism. Training in "modern subjects" such as engineering, ballistics, and communications was "specialized, technical, and nonhumanistic," Edward Shils wrote.[113] Therefore, the "military officer class" represented a "disproportionately larger sector of the modern intelligentsia" and became "major representatives of modernity in technology and administration" in new states.[114] In the "underdeveloped society," Lucian Pye proposed, "the good soldier is also to some degree a modernized man."[115] Further, "the armies created by colonial administration and by the newly emergent countries have been consistently among the most modernized institutions in their societies."[116] The military establishment "comes as close as any human organization can to the ideal type for an industrialized secularized enterprise."[117] Describing the 1952 Joint American Military Mission for

Aid to Turkey, Lerner and Robinson argued that the Turkish army did more than acquire military and technical skills in "the maintenance and operation of modern machinery."[118] The "thousands of young Turkish farm lads" who went through the program, they argued, "acquired a new personality. . . . Along with the physical and social mobility opened to them through the military training program, they acquired also the habits of *psychic mobility*."[119] From then on, "the young discharged soldier is likely to find his traditional society inadequate to sustain his new level of expectation. . . . Hence, it is likely that the soldier returning to his native village will resist falling back into the premechanical era."[120] The Turkish army became "a major agency of social change precisely because it spread among this key sector of the population a sense of identity."[121] Modernity was partly economic, partly institutional, but largely mental, and modernization theorists meant to surpass the prior tradition of accepting the military as an avenue for modernizing institutions and advocate military assistance programs to modernize people.[122]

The Turkish case demonstrates how American scholars in the late 1950s absorbed the prior appreciation for the national military's modernizing effects and transformed them into case studies for modernization theory. Where prior scholars had recognized the political union between the old Ottoman army and the caliph, as if "the Ottoman government had been an army before it was anything else," Lerner argued that Ataturk demonstrated how to modernize the Turkish military while breaking the cycle of military intervention in Turkish politics.[123] Instead, "the maintenance of civilian supremacy in republican Turkey has been a historical fact of the first order."[124] Ataturk broke the cycle of military intervention, which was inherited from classical Ottoman and Persian lore, but which had remained "a regular feature of the postwar scene." Turkey offered a model, then, in which the military, "historically the chosen instrument for imposing stability, can be transformed into an agency that promotes mobility within a framework of civilian ingenuity and innovation."[125]

Even after the military overthrew the civilian government in 1960, Lerner and Robinson still praised the Turkish army. *World Politics* published their 1959 presentation as "Swords and Ploughshares: The Turkish Army as a Modernizing Force" in October 1960, with a new ending that excused the May 1960 coup as over-eager haste from "frustrated" military leaders from one of "the new nations-in-a-hurry."[126] When Turkey's civilian government refused to lead those "major programs that shaped national development—basic education, technical training, [and] industrial production," Lerner and Robinson argued the military could be counted on to intervene.[127] Turkey also seemed to demonstrate that military intervention in politics was not inevitable—or, at least, it did not have to be disastrous.[128] Reinterpreting recent Turkish history as a modernization success story had enormous strategic significance for modernization theorists–turned–policy makers who believed they could achieve similar results in other contested areas.

For a few years after 1959, Americans sought other examples where military rulers also triumphed as modernizers. In 1961, UCLA's Amin Banani argued that Reza Shah's Iran was a triumph of secular nationalism. The unified, standing army permitted the shah to undermine traditional clerical power, reform the judicial and educational systems, and set Iran on the path to economic liberalism.[129] In Egypt, where historians attributed the rise of nationalism in the nineteenth century to the humiliating experience of military defeat, a new CIS-sponsored study by Princeton University sociologist Morroe Berger reevaluated the importance of Western tutelage on the Egyptian military under Ali. Using the typical lexicon of modernization theory, Berger argued that for Mehmed to create a modern, self-sufficient, and efficient army, he needed "new educational methods, a new kind of industrial discipline, a different sort of administrative arm with new and unfamiliar loyalties—in short, virtually a new society."[130] Mehmed failed, Berger wrote, because he either failed or refused "to invoke the symbols of nationalism or these symbols were beyond his own imagination."[131] The Egyptian military's great modernizing power receded after Mehmed died in 1849, revived after the British occupation in 1881, but peaked after World War II: "Nationalism, and political and social reform—*in a word, modernization*: this was the main preoccupation of many of the younger officers reared in the period of British imperial decline."[132] In Taiwan, a partnership between liberal developmentalists in the State Department and the Economic Cooperation Administration appeared to have harmonized US government agencies, Taiwanese government intervention, and private investment to produce "military Keynesianism" in the shadow of communist China.[133] In all three places, harnessing and projecting nationalism remained the modern military's major function, and nationalism was one precondition for modernization.[134]

Nationalism formed the quintessential aspect of the modern military identity. For modernization theorists, training in citizenship was the most significant feature of the modern army's process of acculturation.[135] The army, just like civilian elites and policy makers, shared the same goals of stability, order, national unity, and rapid modernization: "These are the goals of the educated African nationalist," Coleman and Brice wrote.[136] From Turkey to Asia to Africa, young men from isolated villages "now suddenly felt themselves to be part of the larger society."[137] The military's "strong sense of nationalism and national identity, with pervasive overtones of xenophobia" threatened (or promised) to "repress tribal and separatist attachments."[138]

Modernization theorists considered valid only those forms of nationalism that, for residents of newly decolonized states, required external reference points that were only available through instruction from Western sources. The same nationalism that resulted from participation in the external, modern military project was supposed to prepare the military man to acknowledge graciously the inadequacy of his own society. When "large numbers of officers are forced to look outside their society for their models," Pye wrote, "they have greater awareness of international standards and a greater sensitivity to weaknesses in [their] own society."[139] Further, military leaders were more self-confident, more able "to deal frankly and cordially" with external authorities, and more willing to accept correction from the West.[140] Modern men could also face facts: "Military leaders are often far less suspicious of the West than civilian leaders because they themselves are more emotionally secure."[141] Therefore, Pye continued, "it is easier for the military leaders to accept the fact that their countries are weak and the West is strong without becoming emotionally disturbed or hostile toward the West." Rostow called this supposed phenomenon "reactive nationalism" and predicted that newly modernized elites would uproot their traditional societies in order to protect them from humiliation by foreigners.[142] This very conviction permeated the American military assistance program that brought military members from dozens of countries to the United States for professional education and training and formed the core of American military assistance policy to Ghana for the next thirty years. Of course, Ghanaian policy makers and Ghanaian officers carried different assumptions about the practical and ideological significance of their training.

Not only did these theorists assume that the habits of the modern military could spread to the rest of society, but by identifying indigenous officers as an alternate postcolonial elite, theorists easily integrated existing American military assistance policy from the early 1960s. Officers especially, Pye wrote, were "spiritually in tune with the intellectuals, students, and those other elements in society most anxious to become part of the modern world."[143] They were also blank slates who did not carry the intellectual baggage of the independence struggle and who might respond eagerly to the supposedly merit-based opportunities the military provided. "An Army officer at present," William Gutteridge wrote about Ghana in 1965, "is more likely to be the son of a peasant cocoa farmer or a post office official than of a professional man, who will probably have educated his son for the bar or the civil service or a similar occupation of established prestige."[144] Those from humble families or rural regions who had secondary education might choose the military to overcome the hindrance their social or regional background posed.[145]

Modernization theorists recognized that military service during the colonial era carried a stigma. Men from repressed minority groups often served, or the uneducated, or those without other economic options.[146] In Ghana, colonial-era soldiers were called "Abongo boys," a degrading epithet for men who became soldiers because they had no education, had dropped out of school, or were "failures in general, who turned to soldiering as a last result."[147] In the 1940s and 1950s, these were mainly Ghanaians from the north. Ghanaian historian Stephen Kojo Addae suggests, "No sensible young man with a secondary education (and these were overwhelmingly southerners) would want to be a soldier."[148] As the protomodern postcolonial institution, the military-turned-modernizer offered, modernization theorists argued, education and social mobility and promised to replace regional ethnic conflict with national pride. Lerner and Robinson claimed the military experience had fully integrated Turkish society: "Even ethnic minorities are now accepted as candidates for officer status. In the fall of 1955, indeed, several *women* were admitted to Turkey's War Academy as cadets."[149]

Applying military assistance in ways that harnessed the power of this "reactive nationalism" required a new approach. The "Turkish experience," Lerner argued, showed that only with deliberate planning could military assistance attain "the maximum effect upon the economic and social development of the recipient country."[150] The nature of American diplomatic service overseas, however, and the limited formal contact between State Department employees and foreign militaries meant that few lessons in modernity via military training could occur there.[151] It was essential that foreign military personnel come to the United States. Therefore, Lerner recommended, "Foreign military personnel coming to this country for professional or technical training should perhaps also be exposed to our economic processes, political institutions, and social practices."[152] Although this type of professional military training "may well come to differ somewhat from that which is most appropriate in the United States or Western Europe," he argued, "perhaps, learning from the Turkish experience, we can help other countries to harness their national motivation to become strong, via military organization, more directly to the drive for accelerated development."[153]

* * *

As the 1950s ended, influential social scientists who articulated a technocratic solution to the political, economic, and cultural poverty of the "developing world" overcame their reluctance to endorse military assistance as a strategy for modernization. They did so partly because the multibillion-dollar Mutual Security Program, flawed as it was, offered a lucrative path to enact their projects. National security still generated public support, while Americans generally quailed at foreign economic aid. Supposed success stories in Turkey, Egypt, and Iran also lent intellectual credibility to the power of military service to transform individuals. The method and the place where military men and women received those lessons in nationalism, technology, professionalism, and modernity became important links connecting modernization theory, American security assistance policy, and programs that brought foreign military personnel to the United States.

## Modernization Theory Becomes Policy

"The Rostows of this world are not content merely to write books," historian Mark Mazower writes, "they aim to shift policy," and modernization theorists ascended to positions of political influence in the new Kennedy administration.[154] "Euphoria reigned," recalled Kennedy's "court historian" Arthur Schlesinger Jr., "we thought for a moment that the world was plastic and the future unlimited."[155] Rostow crowed, "Since Nov. 8, 1960, [Kennedy's election], a most extraordinary concentration of thought and memoranda-writing has taken place on the question of how to reorganize foreign aid."[156] At the CIS, Millikan and Lerner continued to mingle social science scholarship with influencing policy. Pye took a position at the State Department.[157] Rostow, who had coined Kennedy's evocative phrases the "New Frontier" and "Decade of Development" during the campaign, became one of Kennedy's most influential advisors. He set right to work dismantling the "old look" of military assistance, calling it "a defensive effort to shore-up weak economies and to buy short-run political and military advantages" to inaugurate "a coordinated Free World effort with enough resources to move forward those nations prepared to mobilize their own resources."[158] The liberal universalism of modernization theorists offered interchangeable, scientifically derived, and measurable paths that could be transplanted onto any society and accelerated, with care. But for it to operate when grafted onto existing military assistance programs, another, less definable, faith in the transformative power of the international military ideal had to exist first.

Under current leadership, it seemed to modernization theorists of the early 1960s, American military officers were incapable or unwilling to perceive the potential for transforming visiting foreign officers under their charge. At least, they were wasting an opportunity. US military forces offered an effective path to development, but military leaders were slow to understand the revolutionary nature of their task, their human subjects' potential as targets for modernization, or the richness of their environments for teaching.[159] In a 1962 article in *World Politics*, Charles Windle and T. R. Vallance of George Washington University's Human Resources Research Office acknowledged a consensus among intellectuals that training foreign military personnel in the United States had considerable potential to transform target societies, but, just like Harlan Cleveland did in *The Overseas Americans*, they lamented that the American military still concentrated too much on teaching "combat."[160] American military trainers lacked the foresight to combine military and nonmilitary capabilities. "Other potential benefits of training, such as improving the political, social, and economic status of foreign countries, are treated as collateral benefits," they moaned, and military administrators only considered these "byproducts" useful "if they can be achieved simultaneously with military benefits at little or no extra expense" (94). Given the tendency of Congress "to prefer defense aid to foreign aid expenditures," Windle and Vallance advocated maintaining the current programs that brought large numbers of the Global South's new alternate elite to the United States for military training. But they also suggested that an ambitious effort to "diversify institutional commitments within the defense establishment" might wrest the social significance of military training from "a heavy investment in the production and use of weaponry to other socially useful activities" (97). The existing infrastructure of military assistance, then, provided a convenient host through which theorists attempted to insinuate a new ideological program aimed at spreading the lessons of modernization to visiting elites from newly decolonized states.

* * *

In 1961, the ideological content of those programs changed dramatically. When Kennedy ascended to the presidency in 1961, he claimed to close the door on the moribund "Decade of Defense," and announced that the 1960s would be a "Decade of Development."[161] Secretary of State Rusk told Kennedy that Congress and the American people were tired of the "old symbols, maladministration and waste, neo-isolationism, and the protectionism produced by the economic distress in our own country."[162] Rusk called for "A fresh, positive aid program, scaled to the requirements, and presented with persistence and boldness."[163] As he introduced his Foreign Assistance Act of 1961 to Congress in March 1961, President John F. Kennedy amplified that message and called for a massive reform of the foreign aid system that was as much ideological as it was practical.[164] More than just the "mutual security" label disappeared from the American foreign aid program in 1961. "I am recommending the replacement of those agencies with a new one," Kennedy told Congress; "A fresh start under new leadership."[165] All aspects of American aid were fair game for reform.

Where modernization theory merged with prior American assumptions about their exceptional communities, the international military education and training program deflected a challenge and secured a new purpose for itself. Great personalities churned American foreign policy in the summer of 1961 that threatened all aspects of American foreign assistance. On May 31, 1961, as he convened the Senate Foreign Relations Committee for hearings on the Foreign Assistance Act of 1961, J. William Fulbright leveled what seemed like a withering criticism against American military training programs. "Apparently," he challenged, "we have trained military forces for the kinds of battles that are not fought. We have trained individuals who have not had the will to protect their freedom."[166] He captured the mood of the Senate, Kennedy, and much of the American public when he criticized the "routine, unimaginative administration of programs," noting "a belief that we have become bogged down in annual expenditure procedures without much rhyme or reason or planning for the future."[167] In that small corner of the massive foreign aid structure in which international military training and education resided, however, a subtle shift in emphasis reinforced the program's ideological foundation for decades.

As the Senate hearings on the Foreign Assistance Act sprawled across June, across town at Foggy Bottom, Under Secretary of State Chester Bowles channeled Fulbright's anger to assert the State Department's new authority over American military assistance programs. On June 14, Bowles sent Secretary of State Rusk a sweeping proposal, outlining a broad reform agenda and recommending improvements for nearly every aspect of the department's involvement with American foreign policy. Bowles interpreted the State Department's scope of authority over security assistance broadly. "Another area of our overseas operations which we should consider most carefully," he told Rusk, "is the thousands of military personnel from foreign countries who are brought each year to the United States under the Military Assistance Program for training by the US military in the use of new weapons and techniques."[168] To Bowles, "these many contacts provide a ready-made opportunity to create a better understanding of our country, its beliefs, and policies."[169] He called for a reevaluation of the program to "improve their general understanding of the United States, its people and its policies."

What Bowles called "a better understanding of our country, its beliefs, and policies," represented an explicit, and new, endorsement of foreign military training in the United States for a completely different purpose than Fulbright recognized. Although Fulbright railed against all American training missions, those overseas as well as those occurring stateside, and Bowles addressed only that subset that occurred where foreign military personnel received training in the United States, Bowles and Rusk clearly expected different outcomes than military competence, the ability to fight the correct battles, or efficiency of administration. This training did not require a specific, national security–directed justification, and its primary lessons had no specific combat-related relevance.[170] Bowles's recommendation clarified prior State Department policy drafts that searched for a new, coherent rationale for American military assistance programs that encompassed more than just "the military posture of the United States."[171] Those programs that brought foreign military personnel to the United States were an ideal place to start.

Following Bowles's directive, in 1963, deputy secretary of defense Roswell Gilpatrick issued an initial policy memorandum to the secretaries of the armed forces and commanders of all military commands establishing "An Informational Program for Foreign Military Trainees and Visitors in the United States."[172] The Informational Program memo established several "informational objectives" that dictated the types of additional classes, visits, and lessons desired to give foreign military trainees "a balanced understanding . . . of United States society, institutions, and ideals."[173] It required all military education facilities that hosted international students to present lessons in these subjects because, Gilpatrick wrote, "Successful attainment of these objectives and full exposure of the foreign trainee to the nonmilitary aspects of American life are considered to be of importance to the military assistance program second only to the strictly military training objectives of that program."[174] Centers for American military training and education, such as the US Army's Command and General Staff College at Fort Leavenworth, Kansas, had hosted international students from allied states for decades, and many had already developed programs for augmenting their courses' military curricula with tours, guest lecturers, or "demonstrations of American life." This memorandum formalized the agenda of those programs across all Department of Defense institutions. In January 1965, the Department of Defense replaced the policy memorandum with a formal directive that dictated the program's intent and content for the next two decades. The directive changed slightly in 1985, adding "human rights" and "women's rights" as areas of emphasis, but the eleven "significant facets of American life" that the directive aimed to display remained.[175]

The "significant facets of American life" the Informational Program identified read like a modernization theory cookbook, "each of which contributes a sound grasp of our society, institutions and ideals."[176] Regarding American government, the Informational Program should "make a point many if not most foreigners misapprehend" about federal, state, and local government accountability and decentralization, with special attention to political parties and the "loyal opposition." To demonstrate the American economy: tours of both big businesses and labor unions; understanding "the diversity of enterprise and ownership in our mixed economy"; the importance of "education, research, and governmental assistance" in improving agricultural productivity; public housing and welfare projects "will be of particular interest to foreign military trainees." To display American society: public schools and universities showed "the connection between education and a responsible citizenry"; the free press—that one American institution "foreigners find most difficult to grasp"; political and civil rights for minority groups "with special reference to the Negro problem"; visits to baseball and football games, golf matches, rodeos, and regattas.[177] Although each American military facility interpreted and applied the Informational Program according to its own circumstances, these categories had little military application. Instead, they exactly matched modernization theorists' model of "a New Deal on an international scale," which considered full employment, an end to poverty, unionism that embraced big business, inclusive governance, and respect for civil liberties the praiseworthy products of American liberal universalism.[178]

Army officials did not distinguish between formal Informational Program activities and unofficial, community-sponsored activities when describing their programs for international officers. At Fort Leavenworth in 1991, for example, Colonel William Eggering, the American officer in charge of international students, described a typical year's schedule: picnic with Leavenworth sponsors; Wolf Creek Nuclear Power Plant tour; tour of Southwestern Bell Telephone Company; tour of TWA airline maintenance facilities in Kansas City; Greater Kansas City People-to-People rodeo; fall and spring food fairs; Military Order of the World Wars luncheon for international officers; Winter Formal ball; Kansas City and Leavenworth Rotary Club functions; tour of Fort Hood, Houston, and NASA in Texas; Washington, DC trip, including Colonial Williamsburg and Norfolk Naval Base; Kansas State Government tour; Missouri riverboat tour; guest speakers, including a state senator, judge, AFL-CIO representative, National Organization for Women leader, and local school officials.[179] The Informational Program funded some of these activities, including the tours of Houston and Washington, DC. Local organizations including the Leavenworth-Lansing Chamber of Commerce, rotary clubs, Officers Wives Club, and the Greater Kansas City People-to-People sponsored others.

Colonel Eggering made no distinction between functions, since they all matched the "commandant's goal" to "assist officers and families in acquiring a balanced understanding of US society, institutions and goals."[180]

As an attempt to graft lessons in liberal universalism embedded in modernization theory onto an existing military assistance program, Military Internationalism supplied the essential ideological bridges between American foreign policy, the Informational Program, and American communities. First, extant international military training programs provided access to thousands of visitors from the Global South each year. By 1960, military training brought more foreign students to the United States than all other exchange programs combined, including between 90 and 130 international officers attending six-month or yearlong courses at Fort Leavenworth's Command and General Staff College annually after 1956.[181] Next, policy makers—including the secretary of state, deputy secretary of defense, and modernization theorists in the Kennedy administration—agreed that understanding the United States' "society, institutions, and ideals" was as important in producing psychological transformations as any supposedly "military" lessons visitors learned. They also agreed that American civilian communities, not military training staff, best demonstrated those ideals and instituted the Informational Program to codify that faith into military policy.

The Informational Program did not invent interaction between military posts and their communities; it formalized existing bands of cooperation that had formed organically. Local sponsor families had been assisting international officers and their families at Fort Leavenworth since 1952, coffee klatches and international language clubs for wives and families began in 1947, and at its very first membership dinner in February 1963, the Greater Kansas City People-to-People branch made cooperating with Fort Leavenworth's international officer program a top priority.[182] The Informational Program's authors recognized existing community attitudes and programs surrounding foreign military personnel in American communities, then attempted to focus the programs' ideological content to match the goals modernization theory prescribed for creating a new elite in target states.

Finally, although the Informational Program enumerated specific political, economic, and cultural lessons, it assumed that civilian communities supporting each activity automatically represented the ideal for the visitor to experience. In some cases, the communities themselves also assumed as much, as they unselfconsciously merged international visitors into their community organizations. In many cases, as when the Greater Kansas City People-to-People branch boasted of being the exemplar branch for People-to-People International worldwide, the international visitors sometimes became community organization's raison d'être.[183] All the Americans in the process, from modernization theorists to policy makers to communities, believed that the experience of life in ordinary American communities would inspire or reveal a new psychological orientation in international military visitors. Military Internationalism, therefore, provided the unspoken catalyst for the ideals of modernization embedded in the Informational Program.

The location of the exchange mattered, and American education and training programs at overseas locations never attempted to match the ideological and psychological impact of American training within the United States. In Latin America, for example, Kennedy's Alliance for Progress was no less committed to applying modernization ideology through economic and security assistance, but most encounters with Latin American military students took place in Latin America, usually at the US Army School of the Americas in the Panama Canal Zone.[184] The US poured five times as much money into international military training and education for Latin America than Africa in the 1960s and trained thirteen times more Latin American students than Africans, but the vast majority of that training occurred at the School of the Americas.[185] Canal Zone schools made no attempt to provide students with broad exposure to American life. All instruction was in Spanish and emphasized "teaching students with a lower educational level certain basic and less sophisticated military subjects," including automotive and radio repair, artillery, infantry, and military police tactics.[186] By the 1970s, counterinsurgency dominated the curriculum.[187] Uruguayan officers later recalled that courses in internal security, urban counterinsurgency, intelligence, and "interrogation by military intelligence" were the most popular.[188] The Canal Zone schools left a legacy of militarism within some Latin American societies that continued after the School of the Americas moved to Fort Benning, Georgia, in 1984. American-sponsored training for Latin American militaries still took place overwhelmingly within Latin America even after the move.[189] Even in Georgia, anthropologist Lesley Gill argues, despite an initial nod to the ideals of the Informational Program, American faculty and off-base communities employed the language of international exchange imperiously to enforce a culture of inequality.[190] "A Latin American soldier," Gill argues, "would have to search hard before he encountered large numbers of white, middle-class citizens who represent the 'real Americans' of military imagery."[191]

The School of the Americas offered a much starker contrast to the less technical, mostly political, objectives of the international military education and training program for selected military elites from West Africa. It reveals a chasm separating the ideologies of international military education and the actual results of that training indicative of the disastrous Cold War legacy of Third World military intervention.[192] Evidence of American support for human rights abusers in Latin America and elsewhere barely dented the ideological shell protecting the International Military Education and Training Program. Evidence and ideology inhabited separate worlds.

* * *

When reforming the Military Assistance Program to encourage stateside military courses to emphasize lessons in the American way of life, Kennedy, Rostow, Rusk, and Gilpatrick tapped into the public's own faith that it already embodied a modern ideal. The new Kennedy administration energized or steamrolled many competing government factions into supporting the significant reforms the Foreign Assistance Act of 1961 enacted. The act, and coincident legislation creating the Peace Corps and the Alliance for Progress, capitalized on Americans' confidence in the useful—and even transformative—power that living among the American people could have on visitors from around the world.[193] Most of all, Kennedy's reforms securely enshrined those programs that brought foreign military personnel to the United States with a narrative that affirmed, despite a dearth of supporting evidence, their transformative potential, their long-term success at forging strategic partnerships, and their cost-effectiveness. Those "significant facets of American life" the Informational Program held up for foreign military visitors reflected a shared faith in a version of military-based internationalism that provided a fertile ground for applying modernization theory in American communities. The Informational Program also survived long after modernization theory became unfashionable.

## Military Modernization's Eclipse

Events in the mid-1960s shattered the truce between apologists for "military modernization" and social scientists with grave misgivings about the rapid expansion of military governments in the Global South. Nationalism, war, and a rash of military coups throughout Africa, Asia, and Latin America forced social scientists and politicians to reevaluate the usefulness of military assistance programs as a whole.[194] Training and education programs for foreign military members in the United States did not end, however. Other parties in the initial bargain, including the recipients themselves, simply changed or even heightened the rhetoric behind Military Internationalism to defend and uphold the United States' foreign military training program. For some states, including Ghana, the patterns of American military assistance based exclusively on international military training and education became firm, and both the type and quantity of American aid changed little after that.

* * *

Social scientists had always struggled to understand the connection between the military's potential as modernizer and its potential as political usurper. By the mid-1960s, some blamed social science itself for failing to take military sociology seriously or evaluate modernization theory critically. Henry Bienen of Princeton University accused "academic opponents of the conception of militaries as modernizing institutions" of abandoning the field between 1960 and 1965.[195] In the decade after 1967, research on the performance of military rule in new states abounded, refuting the previous assertion that the Turkish experience was an appropriate model for understanding Asian or African states' modernization and pointing to evidence that military officers often were equally corrupt, unable to build legitimate political institutions, and had fewer public administration skills than the civilian governments they replaced.[196] Nevertheless, influential new academic currents also acknowledged the military's potential to maintain order, despite its seeming inability to restore constitutional rule, and tempered Western social scientists' condemnation of military regimes.[197]

In 1965, Samuel P. Huntington forged a new path to understanding, even tolerating, military regimes. Huntington enabled the intellectual detachment of the large and ongoing Military Assistance Program from the ideological sustenance it had enjoyed in the early years of the decade and that appeared increasingly threatened as the Vietnam War expanded. Arguing in *World Politics* that "rapid modernization" often produced political decay, not political development, Huntington suggested that American policy ought to be to *delay* modernization where adequate political institutions did not also exist.[198] Military officers in juntas or dictatorships may be "modernizers par excellence," he argued, but they were frequently "indifferent or hostile to the needs of political institution–building."[199] On the other hand, military officers, as patriotic conservatives, could also inhibit the unregulated expansion of mass participation in government that threatened developing societies. Huntington argued that "modernity" did not entail prosperity, liberty, or progress; those resulted from a just and stable political order.[200] To him, modernization theory was really only thinly disguised universalism. It revealed more about American liberalism than it helped new states achieve stability or prosperity.[201] He was not the only theorist who called for a reevaluation of American foreign policy in the Global South based on a pragmatic acceptance of the reality and potential of military intervention, but he was among the most influential.

By the early 1970s, the catastrophic Vietnam War, the perceived failure of the War on Poverty, the persistence of postcolonial nationalism, and the emergence of radical critiques of the American empire diminished modernization theory as a guide for American policy.[202] Noam Chomsky condemned Kennedy's clique of public intellectuals and technocrats whose "will to power, cloaking itself in idealism" only spread tragedy in the Third World by expanding American imperial power.[203] Andre Gunder Frank argued in 1966 that underdevelopment in the Third World was actually a built-in and intentional counterpart to capitalist development in the First World, while Gabriel and Joyce Kolko revived William A. Williams' criticism of American foreign policy for expanding the American economic empire at the Third World's expense.[204] Critics from the right, such as P. T. Bauer in his 1972 *Dissent on Development*, questioned whether foreign aid did more harm by concentrating power in the inefficient and unaccountable state.[205] Motivated by ideological criticism from left and right, Congress slashed all foreign aid programs after 1967.[206] Thomas E. Morgan (D-PA) of the House Committee on Foreign Affairs captured the mood of a stingy and suspicious Congress, arguing in 1969, "the construction of powerplants, highways, irrigation systems, and even the financing of projects for the improvement of the health and diet of the population frequently do not bring about the basic changes in social, political, and economic institutions which are essential to economic development and the attainment of social justice."[207] The era of grand projects by the "New Mandarins" and the "Economists with Guns" seemed to end, even if Rostow refused to repent of it.[208]

Congress slashed military assistance budgets after 1967 but left international military education and training programs mostly intact. The original faith, embedded in the Informational Program's assumption that life among American communities offered essential political and social lessons and transformed visiting military personnel, preserved them. Writing in *Foreign Affairs*, the House Foreign Affairs Committee's Donald M. Fraser (D-MN) defended the reorientation of American foreign aid that dispatched with the illusion of massive American economic aid creating stable democracies in the Global South and instead "put social and political evolution as the first concern, with economic aid playing the supporting role rather than the other way around."[209] As one new direction for foreign aid, Fraser suggested "providing students from developing nations with a more profound understanding of development processes and their relationship to ideology would be most productive."[210] As if taking their cue from Fraser, with austerity looming over American military assistance budgets, programs that targeted individual recipients of American versions of modernity never lost their ideological luster.

In January 1968, as one bad year ended and another just began, the president, Secretary of State Rusk, and Defense Secretary McNamara planned for further dramatic cuts in foreign assistance by reiterating their support for programs that brought individuals from potential allies in the Global South to the United States for military training. McNamara foresaw no end to the Vietnam War, feared direct communist assaults elsewhere in Asia, and expected the Soviet Union to attempt to displace or replace US or Western influence everywhere else.[211] Despite a 33 percent reduction in Military Assistance Program funding in fiscal year 1968 and further reductions in fiscal year 1969, he refused to alter the overall program's major objectives of providing "friendly countries" with arms, obtaining US access to bases abroad, helping "friendly countries protect the fabric of their societies against internal violence," and "dispos[ing] nations favorably toward the US in their diplomacy [and] public sentiment."[212] All those objectives remained valid, he said, and none was inherently more important than any other, but he calculated that American military assistance only directly contained communist military aggression in Korea, Taiwan, Greece, Turkey, and Iran.[213] Even against internal instability, McNamara reasoned, "there are distinct limits to the usefulness of military assistance."[214]

In all other cases, the goal of American military assistance was essentially political in nature, intended to foster favorable attitudes toward the United States and its policies, especially "in countries where the military is an important political force."[215] Joint training programs, McNamara wrote, "particularly training in the US, [are] the ingredient for which I do not believe there is a real substitute." "Such training is an effective augmentation of our diplomacy," he argued, "and should be continued."[216] Especially in states dominated by their militaries, McNamara argued, even the smallest programs for military assistance provided "a basis for frequent communications between US and local military leaders" that was otherwise "not easily accessible to influence through normal diplomatic intercourse."[217] This was partly an expression of distrust for diplomats. More likely, McNamara idealized the free communication between military officials of similar rank between different states an essential facilitator for Military Internationalism. As a result, McNamara and his successors devoted a larger percentage of shrinking American military assistance budgets to those programs as the decade closed. Up from 6 percent of the total budget in 1962, international education and training absorbed 10 percent of the total Military Assistance Program budget of $800 million in 1967.[218] By 1969, Congress appropriated $375 million for the entire Military Assistance Program (down from $1.6 billion in 1962), but international education and training absorbed 14 percent of it.[219]

In academic and policy circles, by 1970 modernization theory had lost relevance, but its embedded assumptions that motivated military education and training programs had not gone away.[220] Despite widespread disillusion with what the historian of modernization theory Nils Gilman calls "the atrocities committed from Iran to China to Vietnam in the name of modernization," some of the specific projects that modernization theorists championed in the early 1960s continued. What saved them were their intellectual justifications that rested on preexisting emotional foundations, and that, as Elizabeth Cobb Hoffmann suggests in the case of the Peace Corps but that applies equally to those Americans who administered the Informational Program, "did more for the helpers than for the helped."[221]

## Conclusion

For three decades after the deputy secretary of defense proclaimed the Informational Program in 1963, neither the specifics nor the assumptions behind the program changed much. Most other aspects of American foreign assistance policy did. That is because the assumptions were older than modernization theory, and they preceded theorists' ascent to political influence in the early 1960s. For a century, as the American empire spread west across the Pacific and south through Latin America, American soldiers, politicians, and scholars had assumed that the United States' military embodied the ideal of modernity and that it spread modernity in its wake. American communities around military facilities that trained foreign soldiers had incorporated them into their economic and social lives for decades and, later surveys speculated, would continue to do so even if the Informational Program formally ceased. But modernization theorists "usually denied any connections with earlier philosophies of history," Michael Latham argues, and they seized the opportunity of Kennedy's election to assert a new ideological mold onto a popular, and growing, program of American assistance around the globe.[222]

International military education and training became a prime target for modernization theorists' ideological makeover in the early 1960s because those programs already reached so many men and women from target states. In the past, Cold War justifications built around arming and training American allies to combat or contain communism overseas motivated the expansion of training and education programs. When a critical mass of academics–turned–policy makers, largely funded by the federal government, definitively broke with prior liberal traditions that suspected or rejected the military's usefulness as social reformer, they found in place a large, but mostly untheorized, infrastructure for international exchange. This chapter explained how influential scholars such as Lucian Pye, Edward Shils, and Walt W. Rostow filled those intellectual gaps with the principles of modernization theory. With proper guidance from Washington, they thought, American military education and training programs could focus and transmit essential lessons in American modernity. The Informational Program that emerged in 1963 read like a master syllabus of modernization theory adapted for ordinary Americans' use.

Abroad, meanwhile, the apostles of high modernization escalated war in Vietnam, fabricated famine in India, and exacerbated political and agricultural crises in Afghanistan and Philippines in the 1960s.[223] Either unwittingly or as canny political stunts, President Johnson and his advisors tried to turn these manufactured disasters into rituals of global leadership, but the academic consensus behind them did not hold.[224] As academics rejected massive American intervention in military-supported modernization projects in the late 1960s, American foreign policy also retrenched. But neither in rhetoric nor in practice did the Informational Program abandon what modernization theorists first assumed it could accomplish: Visiting military personnel from target countries received from American communities lessons in American democracy and economy to become modern men and women. They internalized these lessons to critique their own societies and better appreciate the United States, first as individuals, and then later on an institutional level as more and more graduates filled their nations' military services. Either under constitutional rule or military rule—it did not much matter which—the American-trained military officers replicated modernity in their home societies, which made them naturally align with the United States in the Cold War.

Since American military education and training programs reached small numbers of military personnel in each target state, and since the transformation took place primarily at a subconscious level, the process was nearly impossible to measure. Indeed, high-profile cases of American-trained military officers periodically going amok occasionally threatened the entire proposition, but the faith held throughout the 1960s. Partly because it was so inexpensive, relative to massive aid projects that failed even more spectacularly, the program remained intact even when Congress slashed all other foreign assistance after 1967. But low cost was built into the ideology. What made the program so persistent was that the specific sites of ideological transfer—American communities—were permanently above question. Indeed, they had always been, even before modernization theory became trendy. Modernization theorists reassessed, approved, and refined the specific lessons in modernity that the Informational Program proposed to transmit but implicitly recognized that the American people behind the institutions it idealized needed little ideological intervention. Such an intervention would have been unlikely to succeed anyway, since the communities had well-established motives for interacting with visiting military personnel and their families. From an American perspective, all parties involved in the interchange—from modernization theorists, to State and Defense Department staff and American military officers at education and training institutions, to civilians around such places—accepted the fundamental goodness of the personal interaction. The Informational Program's faith in the power that ordinary Americans already possessed reveals that modernization was only a temporary and parallel explanation for the larger processes of internationalism already at work in American communities. Grand national strategies, foreign policy objectives, and high-level diplomacy often intervened in the program's global structure, but individual human contact remained above discussion as the essential catalyst for the exchange's fundamental usefulness.

# 2 Independent and International

## *Ghana's National Military and American Assistance Policy*

### Three Boxes

"On occasions such as this," Charles de Graft-Dickson, the Minister Responsible for Defense, told the gathering at Ghana Military Academy, "it is always better not to make long speeches."[1] Canada had donated about "70 odd books and pamphlets" to Ghana Military Academy, which was just a month away from graduating its first class of officers. De Graft-Dickson made sure to mention the eight Canadian instructors teaching at the academy on this August afternoon in 1961.[2] Then, on behalf of the *Osagyefo*, Ghana's president, he thanked the government and people of Canada, and the "short, but very impressive" ceremony ended.[3]

The next day, Lieutenant Colonel Harry Coates, the US Army attaché to Ghana, sent a letter to the Ministry of Defense offering to donate another collection of books and pamphlets—this one gathered from the United States' service academies in New York, Maryland, and Colorado—to Ghana Military Academy.[4] Francis H. Russell, the US ambassador, intended to present the gift personally, so Lieutenant Colonel Coates requested: "A suitable ceremony be arranged," and he "recommended" that Ghana's minister of defense and the chief of defense staff attend.[5] Coates and the minister of defense scheduled the ceremony for September 5, 1961.

Something came up—or rather, something went down—and the Ghanaians postponed the ceremony. On September 1, the twin-engine airplane carrying Brigadier Joe Michel crashed at a Ghanaian air base. He died from burns the next day.[6] Michel was Ghana's second commissioned officer, one of only two Ghanaian generals, the commander and hero of Ghana's brigade in the Congo. He had just been promoted to chief of staff of the *Opération des Nations Unies au Congo* (ONUC—United Nations Operation in the Congo), the highest position any African had attained in an international peacekeeping mission.[7] Ghana's President Nkrumah blamed the crash on colonial powers trying to sabotage Ghana's peacekeeping mission, claiming they were desperate to prevent a Ghanaian from demonstrating that Africans were capable of commanding international troops.[8] In deference, the United States agreed to postpone its book presentation ceremony until after Michel's funeral service.

De Graft-Dickson apparently did not consider this second book ceremony the kind of occasion to avoid a long speech. Neither did the US ambassador, who praised the gift of books as "symbolic of the bond of friendship" between the American and Ghanaian military academies.[9] "There is a natural fraternity among those whose lives are dedicated to preparing for the defense of justice, liberty, and freedom," Ambassador Russell claimed as he praised the particular quality of this batch of books. "I was struck by the titles of these 300 volumes," he said. "They seem to me to be fully as symbolic of the gift itself." Mixed among titles on ordinance, tactics, science, and math he also found "history, government, law, international relations, economics—and one called 'modern civilization.'" Struck, but not surprised, Russell reflected that a text on "modern civilization" fit perfectly on the Ghana Military Academy's library shelves, dedicated as it was to "the preservation of the vast total of human values which that title comprehends."[10]

Just one month later, while on a visit to Moscow, the Ghana Military Academy commandant received another box of books, courtesy of the Red Army.[11] The commandant and Ghana's ambassador addressed the parcel of books to the minister of defense and put it on an airplane for Accra. Nobody saw the books again. Through the spring and summer, the humiliated commandant, chief of defense staff, and minister of defense searched in vain for the boxes. After a year of searching, they quit.[12]

The story of these three parcels of books weaves assumptions about modernity and modernization, postcolonial African nationalism and pan-Africanism, and Cold War politics in Ghana. Political and military leaders in Ghana, the United States, and elsewhere publicly affirmed a faith in international military cooperation, especially in exchanges of education and training, to transcend or smooth the coarse diplomatic relationship between the states. Those exchanges never fully escaped their political context, but they routinely appealed to nonpolitical or transnational ideals of modernity through civil-military relationships that elites of both states accessed for their separate purposes.

From their earliest attempts to strengthen functioning national bureaucracies after independence in 1957, leaders in Ghana's government and armed forces sought international assistance in the form of financial aid, opportunities for education and training abroad, and outside investment in Ghana—both economic and in human resources.[13] The source of that international assistance indicated broad trends in Ghanaian Cold War alignment, to some extent, but also reflected internal competition between factions of the Ghanaian state.[14] Before his ouster in 1966, President Kwame Nkrumah tried to balance the legacy of British organization and training in the army by accepting training and equipment from Eastern Europe and the Soviet Union.[15] He did so partly to assert Ghanaian nonalignment and exert diplomatic leverage against both blocs, partly to weaken the ideological coherence of a military he suspected was plotting his overthrow, and partly to finance his own personal President's Own Guard Regiment.[16] But Nkrumah and his successors never questioned the importance of maintaining at least a token national military, so they also consumed international military assistance because Ghana largely lacked the capacity to produce and cultivate personnel capable of performing military tasks.[17]

Ghana was never simply a client for international military aid. Ghanaian officers *did* receive commissioning training and staff college training in a half dozen countries after 1957, but the very same professional training Ghanaians received at military schools abroad, Nkrumah and his successors also exported to other African states. Nkrumah seized the opportunity to provide troops and support for United Nations operations in the Congo within days of the crisis and maintained his commitment to the mission despite widespread condemnation from other African leaders.[18] Even during periods when Ghana struggled to "feed itself," Ghanaian leaders considered Ghana's capacity to export military expertise central to their pan-African vision for collective security.[19] Other African states sent cadets to Ghana's military academy within three years of its opening in 1961 and to Ghana's staff college in just the second year of its existence.[20]

This chapter argues that for Ghanaians, just as for Americans, the social value of lessons in military leadership, management of modern state institutions, and the vague category of professionalism transcended the national location where those lessons occurred. Ghanaian politicians and military officers—who were sometimes the same—widely and readily reinterpreted international discourses on civil-military relationships to devise their own understanding of how international military resources could serve Ghana's national interests. They consumed those lessons from a wide variety of education and training facilities abroad, with a broad tolerance for the political legitimacy of the sponsoring states. Likewise, as soon as it attained an institutional capacity to do so, Ghana retransmitted those same lessons to personnel from other African militaries—also with a considerable tolerance for the uncertain legitimacy of recipient regimes. Just as the Americans did, Ghanaians acknowledged a wide range of political configurations among partners in international military education and training because they believed the program itself was the mechanism to bring democracy, justice, and even legitimacy to new states. Later chapters will show that the subnational and corporate benefits that flowed from international education made individual Ghanaians and their families reluctant to condemn sponsoring states' political or social virtue.

For three decades after 1960, the United States maintained a small international military education and training exchange with Ghana that bore the full ideological load of American assistance policy there. It was the only military assistance the United States provided to Ghana, despite occasional assurances that the United States considered Ghana an essential, yet undeveloped, potential ally in Africa.[21] Continuously since the early 1960s, American policy makers expected the small military education and training program there to yield long-term American influence, with great potential for future results. The rise of Ghanaian Colonel Ignatius Acheampong in 1972 should have represented a crowning achievement of American military education, but instead he demonstrated its potential downside.[22] Acheampong was a Command and General Staff College graduate and precisely the sort of ascendant military officer the American program proposed to educate, but, when he overthrew Ghana's second republic, the United States mostly disregarded his American education.[23] Acheampong showed that for the United States, Military Internationalism worked best in the abstract.

Ghanaians never rejected American grants for military education abroad, but they responded to the limited opportunities for such training in the 1970s by building their own staff college and accepting officers from other African states there. The Ghana Armed Forces' effort to "Ghanaianize" the staff college demonstrated the fluidity of roles between consumer of international military education and provider. By the late 1980s, a Ugandan military officer attending Ghana Armed Forces Command and Staff College likened the college's annual journal to a mouthpiece for African independence from the "Colonised Mind."[24] By then, Ghana continued to accept grant aid to send promising military officers to staff colleges around the world, but through its own staff college offered professional education and training that Major Tumukunde of Uganda equated with "surviving on our own," as the "proper utilization of the natural and human resources within our countries."[25] The irony was that little of Ghana Armed Forces Command and Staff College was exclusively African. But Military Internationalism could bind Ugandan officers to ideal African communities just as easily as it bound Ghanaian officers to American communities.

## Training and Education, Integration and Isolation: US Assistance Policy to Ghana before 1966

Ghana epitomized most of the preoccupations, ambitions, and limitations of American foreign policy with respect to decolonizing Africa in the early 1960s, and sponsoring education and training for Ghanaian military personnel represented the standard American military assistance response. Proxy battles between the superpowers in the Third World may have been the Cold War's main event, as Odd Arne Westad argues, but lingering colonial influences complicated both American and Soviet neocolonial strategies for engaging newly independent states.[26] Leaders of African states resisted the hegemony of the Cold War powers as strenuously as they battled colonialism, and American foreign policy makers knew it.[27] Ongoing deference to colonial powers limited the United States' relationship with new states, making them "triangular," while at the same time opening a "window of vulnerability" for the Soviet Union to attack continued Western hegemony over the Third World.[28] Therefore, as Rostow recalled, "The essential gamble of the United States in the developing world of the 1960s was on the ultimate strength of the determination of nations to maintain their independence."[29]

Ghana provided modernization theorists an opportunity to test whether they could assist newly decolonized nations' development without threatening their independence. American economic assistance to Ghana in the early 1960s revolved around the Volta River project: a massive dam, hydropower plant, and bauxite aluminum processing plant, built by a consortium of American manufacturers under the Kaiser Aluminum Company and financed largely through US government–backed loans.[30] By most accounts, the dam was a modernization success story, providing Ghana fresh water and enough surplus electricity to export to neighbors since 1966.[31] But throughout 1961, fear that Nkrumah would nationalize the dam and aluminum industry and fear of repeating Eisenhower's blunder of reneging on a promise to fund the Aswan Dam in Egypt caused President Kennedy and his advisors to vacillate over extending government insurance for Kaiser's investment.[32] By September, Kennedy privately decided to abandon the plan, and "he had given up" on Nkrumah. Kennedy reflected that he "wouldn't mind giving them a little aid," but thought Nkrumah was being "unnecessarily difficult with us . . . as bad as he's been in Belgrade, as well as his desire to send 400 soldiers to Russia."[33] But with the help of Secretary of State Rusk and Under Secretary of State Ball, who feared that governments across Africa would accuse the United States of applying "political strings" to aid that were "indifferent to Africa's development needs" and that "refusal to go forward here were be misunderstood all over Africa," Kennedy changed his mind.[34]

Fear that the Soviet Union would take over the project, Kennedy's claim to have deep personal relationships with so many African nationalist leaders (including Nkrumah), and fear that Nkrumah would expel the first group of Peace Corps volunteers he had welcomed with great fanfare earlier in the summer convinced Kennedy to finance the dam.[35] Most of all, Rostow recalled, Ghana "provided an occasion to test and explore the complexities of the doctrine that long run rather than a short run American political interests should guide aid allocations."[36] In Ghana in 1961, modernization theorists prevailed, supporting a long-term development program that integrated private industry, government subsidies, and high-profile technological projects with modernization ideology, underwritten by faith in its power to knit psychological allegiance with the United States to a massive infrastructural project. The Volta River Dam also represented a point of intersection between the parallel tracks of decolonization and the Cold War. All aspects of American economic, political, and cultural diplomacy interacted in this decision. It involved not only social scientists and policy makers in capital cities but also American Peace Corps volunteers in rural Ghana, Ghanaian army cadets in the Soviet Union, industrialists, and diplomats across Africa and Asia. The dam represented a high-profile example of American development in Ghana, but behind that major project, we find the intersection of several other currents of American foreign policy with Ghana. Whether the United States should provide military assistance to Ghana was one such current.

In the spring of 1963, the Office of the Director of Military Assistance within the US Department of Defense issued a five-year plan providing the rationale, objectives, and details for the United States' military assistance policy to Ghana. The plan acknowledged that the United States was just one among several states from which the Ghanaian government sought economic or military assistance. While the United States recognized Ghana's ongoing military relationship with the United Kingdom, American policy makers also perceived that the nonaligned Ghanaian government carefully considered the ideological implications of any military assistance it received in order to maintain its anticolonial, nonaligned reputation. Therefore, the United States selected a military assistance policy that deferred to the United Kingdom and Canada to assist with training Ghanaian officers in Ghana but would augment via grant aid advanced or specialized military education and training in the United States.

The United States' military assistance plan for Ghana in the early 1960s typified the new faith, inspired by modernization theory, that a small international military education and training program could yield great improvements in any contested state's political and economic development, augment or eventually replace the former colonial power's influence, and secure new states' long-term ideological orientation to the West. The American plan predicted great rewards from a small investment. Following guidance from Kennedy's policy planning staff for non-NATO recipients with "no demonstrable military requirements for assistance," the American plan recommended using as little military assistance as possible, applied "primarily for the achievement of political purposes."[37] It sought to supplement, rather than supplant, British and Canadian military assistance efforts in Ghana, to assist in "civic action projects which will contribute to social and economic development," prevent or limit Soviet "bloc penetration" in Ghana, and foster an "anti-Communist, Western oriented military community."[38] "The British and Canadian governments have primary responsibility for providing military assistance to the Ghana Armed Forces," the plan acknowledged, therefore US policy would be limited to training Ghanaian personnel at service schools in the United States, which the Ghanaian government eagerly accepted in the past.[39]

These objectives, and the specific programs the United States employed to achieve them, reveal that American policy makers understood the United Kingdom's ongoing, but declining, influence in its former colonies, and Ghana's tense relationship with its former colonizer. In the educational exchange program they advocated, the Americans also revealed faith in the ability of international military education and training to create significant psychological effects, which they assumed would bring about Ghana's strategic realignment toward the United States. What is most interesting is that American planners assumed that the programs that would achieve these results required little administrative oversight. Their results would flow naturally from the nature of the experience of international education and training itself.

American planners in 1962 considered military assistance from Canada an acceptable proxy for American or British support, but integrating the three nations' programs in Ghana also revealed the problem of finding the proper balance for American intervention versus deference to former colonizers in Africa.[40] Some American attempts at assertiveness were just clumsy, as in March 1962 when the State Department proposed a sweeping scholarship plan in the United States for military elites from across West Africa. The French secretary of state for African-Levant affairs, Jean Sauvagnargues, ridiculed the idea.[41] "There are many fields in Africa where American aid and training [are] badly needed," he informed US Ambassador James Gavin, "military training is not one of them."[42] On one hand, Sauvagnargues's colorful and vehement opposition to the American proposal reveals one aspect of the complicated "triangular" relationship between American, European, and African states at this time.[43] It also reveals another instance where American planners expressed faith in the universal modernizing power of educating and training elites from the Global South. The concept was of joint State-Defense Department origin; it intended to engage a wide variety of African nations despite language difficulties; it assumed that military education in the West could impart lessons or an identity that greatly outweighed any differences in national tactical or organizational models; and it represented American faith that the education of a few individuals abroad could transform vast segments of African society. These assumptions weave several aspects of the state-endorsed version of internationalism the United States applied with more tangible results in Ghana.

The Africa-wide training proposal went nowhere, but the US director of military assistance made training and education the cornerstone of American assistance policy with Ghana in 1962 and 1963. The US plan deferred to the British and Canadian governments and "deliberately restricted" the US program to "training Ghanaian officers and enlisted personnel" in service schools in the United States.[44] But while the United States' strategy incorporated British and Canadian participation, the Americans were also readily prepared to supplement their efforts when British, Canadian, or Ghanaian resources ran short. They had once already, in March 1962, when "Ghana was forced to cancel a sizable number of officers and enlisted men from attending courses of training in foreign countries other than the US due to lack of funds."[45] Offering military education gave the United States a method to enlarge its participation in the "triangular" relationship without directly interfering in the former colony.

The United States' appraisal of Ghana's foreign policy objectives and domestic politics also encouraged military assistance based on foreign training alone. American planners expected Ghana's economy to be unable to support Nkrumah's desire to expand the armed forces to include a jet fighter squadron, an additional artillery regiment, and new warships.[46] For the fiscal year 1963–1964, the Department of Defense projected military expenditures of $40.7 million to absorb 11 percent of Ghana's national budget.[47] Even though American planners expected the Soviet bloc to continue offering equipment and training to Ghana, the United States refused to sell or loan expensive or complicated military equipment. "Since arms and equipment for the military must be considered non-productive in nature," the plan stated, "insistence on such purchases will inevitably detract from Ghana's ability to further general economic development."[48] Too large a military assistance program "would be embarrassing to the Ghana Government since its propaganda, with respect to the US position in the NATO Alliance and its attempts to label the US as 'Neo-Colonialist' in its relations with emerging African countries, would be hypocritical and ineffective."[49] Despite "some expressed uneasiness on the part of left-wing elements in the Government who feel that the Ghanaians involved are being subverted in the US while in training," Ghana "eagerly accepted" American grants for training in the United States.[50] The director of military assistance staff weighed Cold War competition over Ghana, lingering British influence there, and the harmful effects of militarizing Ghana's economy, and they reaffirmed international education and training as the most effective and cost-effective path for American security assistance to Ghana.

American planners understood how damaging dependence on foreign suppliers for military equipment and arms transfers could be, but they placed great faith in the psychological potential of their alternatives to technical military assistance. The American program fostered "a spirit of Western oriented democratic attitudes in the military forces of the newly emerging nations in Africa."[51] To the United States, "the Ghana Armed Forces are considered apolitical," but when the twenty-one Ghanaians who attended courses in the United States between 1962 and 1963 returned to Ghana, they came back "enthusiastic about US military training techniques, facilities, amenities, arms, equipment, technical skills, etc."[52] Even more, they developed "an appreciation for the abundant life of Americans."[53] "This enthusiasm has spread throughout the Ghana Armed Forces," the assistance plan projected, and "the result has been a marked friendliness and eagerness for information about the US military affairs and American life on the part of the personnel of the Ghana Armed Forces."[54] New initiatives in the US Military Assistance Program—which became the Informational Program later in 1963—promised that "visits to such places as American homes, local industries, industrial and cultural exhibits, country and state fairs, places of historical interest, and sporting events, will be effective in fostering goodwill and, acquainting the individual with American culture."[55] American planners were convinced that this "surprising enthusiasm" was responsible for the fear that those "left-wing elements" in the Ghanaian government had that Ghanaians were "being subverted in the US."[56]

The 1963 American military assistance plan for Ghana bonded modernization theory's assumptions about the transformative potential of intervention among news states' elites to American military assistance policy in Africa. The plan proposed no equipment sales and recommended neither sending military assistance teams nor sponsoring any civic action projects within Ghana. It advocated only "providing technical and administrative training of such a nature as not to materially increase the combat effectiveness of the Ghana Army or generate requests for US equipment."[57] Grant-based training aid for Ghana fit within the United States' larger military assistance objectives in 1963, which supported "any action on the part of the US that might influence [an] individual or group of individuals to a better understanding of the basic precepts and culture of the American system."[58] The United States planned to provide between $140,000 and $160,000 in grants, which would enable twelve to fifteen Ghanaian military personnel to receive training or education in the United States each year.[59] American planners assumed this level of support sufficient to help create "a well-trained, experienced and disciplined Officer and Non-Commissioned Officer Corps," to "foster a spirit of Western oriented democratic attitudes," and to "counter unfriendly or hostile propaganda" emanating from the Soviet bloc.[60] The plan considered military assistance a multinational endeavor. It acknowledged the primacy of the British Commonwealth in specific assistance programs but also incorporated the variety of national influences over individual Ghanaian trainees within a larger perspective that encompassed all international encounters as equally influential, and equally valid, toward transforming the Ghanaian armed forces into a modern institution within a modernizing state.

This plan acknowledged the United States' limited immediate influence in Ghana but also revealed great confidence in the long-term, nonspecific value of international military travel to reap strategic benefits. The United States' assistance plan for Ghana also expressed precisely the political, psychological, and economic rationale for the International Military Training and Education Program that became the centerpiece of American policy toward sub-Saharan Africa in general. American military assistance planners were almost delusional about the positive influence of a visit to the United States, even suggesting that the United States should restrain itself to keep the Ghana government from seeking to counter America's rapidly expanding influence with more Soviet aid. By 1963, the US Military Assistance Program had evolved, acknowledging that the greatest benefit of these visits to the United States came not from the military training itself, but through the psychological and social transformation encouraged through the Informational Program. Finally, with complete faith in the political windfall of these psychological effects, the United States created an enduring mold for American military assistance policy to Ghana that would last a quarter century.

## Ghanaian Nationalism, Pan-Africanism, and Military Internationalism in the Early 1960s

In the early 1960s, both British-educated Ghanaian military leaders and pan-Africanist, anticolonial politicians under Kwame Nkrumah already incorporated international political, economic, military, and cultural resources for their own benefit. When integrating Ghana into their larger security and development objectives in Africa, American diplomats and military assistance planners merely tapped into lively, preexisting discourses over the relationship between Ghana's president, military hierarchy, the extensive use of training programs from foreign countries for Ghanaian personnel, and the presence of foreign military training professionals in Ghana. These churning local and international influences complicated even the most capacious understanding of postcolonial nationalism in Ghana. To the extent that so many contemporary standards for military organization and training had non-African—often colonial—origins, one might expect to find the Ghana Armed Forces seeking equipment and assistance from abroad. But from the earliest days of independence, competing Ghanaian political and military leaders alike employed internationally developed descriptions of professionalism and loyalty to frame their own discourses on the nature of African self-reliance and the duty of the national military in dangerous political times.

Just as American policy makers did, Ghanaian leaders assumed professional military training created common, universal products. Elites in both states expected that the ideals of professionalism, service, deference to civilian rule, and loyalty to result regardless of the national origin of the training. Speaking at Ghana Military Academy in 1962, Nkrumah praised loyalty above other martial virtues, but he and the British-trained army brass who increasingly opposed him had different ideas about the part of the Ghanaian state to which that loyalty should adhere. Both approached the Ghanaian nation with different definitions, which gave the specifically international programs they both depended on to expand and mature Ghana's young military different ideological significances. The 1966 coup that deposed Nkrumah was, in one sense, a culmination of the tension between powerful domestic political blocs whose own visions for the nation could only be realized through extraordinary international intervention and assistance. In this conflict, the international and the local interacted directly, often at the expense of the national.

The epidemic of African coups in the mid-1960s set off waves of scholarship in the West attempting to explain the military's ascendency among competing domestic factions in postcolonial Africa.[61] After Ghana's coup in February 1966, scholars plumbed the relationships between the charismatic Nkrumah and his ethnically disunited state, or the Western-oriented military and Nkrumah's socialist Convention People's Party, or the struggle between ex-colonial and neocolonial powers and new states in Africa.[62] In their memoirs, which followed shortly after the coup, both Nkrumah and the military officers who overthrew him described the events leading to the coup in terms of national honor, or in pan-African terms, or in terms of economic or political independence from either Western or Eastern ideological blocs.[63] Although some scholars suggested examining the sociology of the armed forces, rather than the national origin of its training, as the most important determinant of military behavior, in most of these accounts, the modern, independent nation-state reigned as the ideal political, economic, and cultural form against which Ghana's experience with personal rule and military rule only deviated.[64] Despite modernization theorists in the United States' early hope that international military training had the power to create the ideal, modern nation, in Ghana only disappointment resulted. Only in the long view did Military Internationalism offer promise as a corporate or national, not just individual, identity.

For its part, the Ghanaian government recognized that military assistance, like all other forms of international cooperation, could serve its own national objectives for nonalignment and development. But the postcolonial context of international competition and Ghana's own limited resources restricted its options to consume this assistance. Before 1966, Ghana under Nkrumah attempted to balance American and Soviet offers of military assistance with Ghana's prior and ongoing attachment to the Commonwealth for military organization and training. Ghana also attempted to engage military expertise from India, Yugoslavia, and Israel while aggressively committing Ghana's military to UN operations in the Congo, as Nkrumah hoped to balance Ghana's nonaligned image with his own pan-African ambition. In Nkrumah's foreign policy, nationalism and internationalism merged.

* * *

Although officers composed only 10 percent of the Ghanaian military after independence, all of Ghana's competing political factions considered recruiting, training, and promoting officers the most important aspect of creating Ghana's national army and one that Ghana could not perform without international assistance.[65] Nkrumah sought to expand the armed forces from 4,000 to 7,000 by 1961, to 14,600 by 1966, and to create a separate air force and navy.[66] In 1960, he unilaterally claimed for Ghana authority over the UN mission to the Congo.[67] Most importantly, he sought to "Ghanaianize" the armed forces, to eliminate its dependence on British officers who were "seconded" (serving on loan) to the Ghanaian army, and to place senior Ghanaian officers in positions of command. A shortage of Ghanaian officers complicated this plan. The British had created only twenty-eight Ghanaian officers between 1953 and 1957. Most graduated from military academies around the commonwealth after a short preparatory course for prospective officer cadets at the British-run Regular Officers' Special Training School facility in Teshie, near Accra.[68] All but one had first been enlisted soldiers in the colonial army, this both a symptom and cause of the low esteem potential elites held for military service.[69] At independence, however, there were no Ghanaian officers in the rank of lieutenant colonel or above.[70] Nkrumah and the Ghanaian Ministry of Defense projected in 1959 that Ghana needed to train 180 more—30 per year—by 1965.[71] Ghana Military Academy partially filled Ghana's urgent need to create Ghanaian military officers to replace British ones and to supervise the expansion of Ghana's military, but even this national effort was extraordinarily international. Ghana Military Academy relied heavily on foreign officers as instructors; meanwhile, Nkrumah dreamed that it would also train officers from across postcolonial Africa.

Producing Ghanaian officers was never simply a Ghanaian endeavor. By Nkrumah's request, the Royal Military Academy at Sandhurst's director of staff, T. J. S. Anderson, visited Ghana in December 1959 and drafted the plan that guided the creation of Ghana Military Academy.[72] In April 1960, Ghana Military Academy opened on the facilities of the former British-run training school in Teshie. To fill Ghana's urgent need for military officers, Ghana Military Academy would have to accept between twenty and forty cadet candidates for each eighteen-month course, beginning in April 1961.[73] A group of incoming cadets formed an "intake," and Ghanaian officers have referred to themselves by military academy intake number ever since.[74] According to Anderson's initial plan, Ghana would still send five officer candidates per intake to Sandhurst and five cadets from each intake to other commissioning sources abroad, including India, Pakistan, Australia, and Canada.[75] Ghana expected one-third of the 180 officers produced between 1961 and 1965 to receive their commissions from a source outside Ghana. To Anderson, it was

**TABLE 2: SUMMARY OF MILITARY ASSISTANCE TO GHANA, 1960–1992**

| Supplier | Year | Type of Assistance | Other Comments |
|---|---|---|---|
| Canada | c. 1960 | Aircraft:<br>14 "Beaver" aircraft<br>12 "Otter" aircraft<br>8 "Caribou" transport aircraft* | |
| | 1962–1968 | 30-person Canadian Armed Forces Training Team*/** | |
| Finland | 1964 | 85 Mortars (various sizes)*** | |
| France | 1975 | 4 Aerospatiale Alouette III helicopters*** | |
| | 1980 | 24 Exocet anti-ship missiles*** | |
| Germany (East) | 1964 | 2 Exchange officers from Ministry of State Security conduct secret service/ intelligence courses in Ghana* | |
| Germany (West) | 1975 | 2 patrol boats*** | |
| | 1978 | 4 Lurssen patrol boats*** | |
| | 1980 | 4 Lurssen patrol boats*** | |
| India | 1963 | Officer cadet training in India** | Unknown numbers |
| | 1987 | 1 Hindustan HTT-34 trainer aircraft*** | |
| Italy | 1966 | 7 Aeromacchi MG-326 light attack aircraft with Italian instructors* | |
| | 1978 | 9 Aeromacchi MB-326K light attack aircraft*** | 4 in 1977, 1978, 1 in 1980 |
| | 1983 | 8 Marchetti SF-260TP trainer aircraft*** | |
| | 1989 | 2 MB-339A trainer aircraft*** | |
| Libya | 1983 | "Unspecified Quantities of Military Equipment, mostly artillery pieces and ammunition, as gifts" * | |
| Netherlands | 1975 | 6 Fokker-VFW F.27 transport aircraft*** | |
| | 1976 | 1 Fokker-VFW F.28 transport aircraft*** | |
| UK | 1960–1971 | 250-person Joint Service Training Team*/** | |
| | 1960 | Aircraft:<br>12 "Chipmunk" training aircraft<br>3 "Heron" transport aircraft<br>9 helicopters* | |
| | 1964 | Vehicles:<br>12 "Ferret" scout cars<br>4 armored cars*** | |

| Supplier | Year | Type of Assistance | Other Comments |
|---|---|---|---|
| | 1964 | 2 Corvette patrol boats (refurbished in 1975)<br>1 naval frigate */*** | |
| | 1974 | Aircraft:<br>6 Skyvan transport aircraft*** | |
| | 1976 | Aircraft:<br>1 Bulldog trainer aircraft*** | |
| USA | 1962–1992 | 673 International military education and training students | |
| | 1989 | 1 C-212 aircraft*** | "Purchased for Ghana Air Force from a Private User" |
| USSR | 1961 | 8 IL-18 transport aircraft* | 4 returned to USSR / 4 transferred to Ghana Airways airline |
| | 1961–1966 | 76 Officer cadets' training in USSR* | |
| | 1963 | Mi-4 helicopter* | |
| | 1964 | Vehicles:<br>10 BTR-152 armored personnel carriers<br>3 tracked assault gun carriers** | |
| | 1965 | Weapons:<br>15 anti-aircraft artillery guns<br>24 light artillery guns<br>20 heavy machine guns* | |
| | 1966 | Soviet crews construct Tamale Air Base* | |

Sources: *Berry; **OASD; ***SIPRI; ****Other. Even in the case of supposed "purchases" of military goods, it is not clear how much, or if, Ghana actually paid for these items. Laverle Bennette Berry, *Ghana: A Country Study, Area Handbook Series* (Washington, DC: Library of Congress Federal Research Division, 1995), 282–90; Office of the Director of Military Assistance, OASD (IS), "Military Assistance Plan: Ghana (Secret)," sec. K, RG 330; Survey of SIPRI yearbooks, beginning with *SIPRI Yearbook of World Armaments and Disarmament 1968/69*. (Stockholm: Stockholm International Peace Research Institute).

the combination of Ghanaian-trained officers and "the present arrangement of sending a number of officer cadets to overseas establishments each year" that offered "every prospect of building up a first-class corps of officers."[76]

Even the value of the officer's commission itself required international validation. Ghanaian officers feared that a military commission earned in eighteen months in Ghana might lack the prestige of a continental military commission. Practical considerations—specifically, Ghana's urgent need for officers—dictated the choice of an eighteen-month course, but so short a course seemed also to require strict entry standards.[77] An eighteen-month course "would produce worthwhile results provided that only candidates with the right sort of entry qualifications as well as the essential qualities of character and mind are accepted," Anderson projected.[78] He noted, "The course at the Ghana Military Academy will be notably shorter than those conducted in similar academies elsewhere," including Sandhurst, the US Military Academy (West Point), the Malayan Federated Military Academy, the Pakistan Military Academy, or the French, Canadian, or Australian academies.[79] In his history of the Ghana Armed Forces, Stephen Kojo Addae suggests the "advantageous international character" of the Sandhurst diploma made local African training seem "parochial" by comparison.[80] On the other hand, Ghanaian officers who graduated from Sandhurst did not accept the "Queen's commission" at ceremonies in England and instead received a commission from Nkrumah upon returning to Ghana.[81]

Such a declaration of independence served several goals. Nkrumah and other pan-Africanists considered maintaining a core of African elites critical to validating their charges against the absurdity of racism that justified paternalist apologies for lingering colonialism.[82] A national army was an essential ingredient for sovereignty, and Nkrumah's ambition for African nations to liberate themselves depended on the capacity to create and maintain one.[83] And as a leader among nonaligned states, Nkrumah shaped his ideology of "positive neutrality" from Cold War manipulation in order to resist "pacts and agreements" that drew small states into superpowers' orbits.[84]

From its inception, however, Ghana Military Academy struggled to hire Ghanaian instructors and depended on international officers to fill essential training posts. As early as 1959, the Ministry of Defense asserted "the aim of Government policy to Ghanaianize the academy teaching services as soon as possible" and requested the Ministry of Education to help find at least four suitable Ghanaian university graduates for the academy faculty prior to its April 1961 opening.[85] Symptomatic of the low esteem of the military in Ghanaian society in 1959, the Ministry of Defense acknowledged difficulty finding four suitable candidates and asked for an all-out effort to find at least one Ghanaian, who could "be closely associated with the administration and formulation of policy in the early stages of the Academy."[86] Meanwhile, expatriate officers from the British army served as instructors at Ghana Military Academy until September 1961, when the Canadian Armed Forces Training Team, Ghana replaced them.[87]

From its inception and initial planning stages and throughout its early years in operation, Ghana Military Academy employed international standards, international staff, and international methods to train officers in Ghana's national army. "Ghanaianization" never implied self-sufficiency. Instead, Nkrumah and the Ministry of Defense concentrated on producing Ghanaian officers, whose very presence projected African authority, competence, and modernity by international standards.

A new kind of modernity became the curriculum. For sixty years prior to Ghana's independence, the Royal West African Frontier Force had applied European racial and cultural categories to identify "martial races" by ethnicity (Hausa from the north) and religion (Muslim) to seek specifically *unmodern* recruits.[88] Colonial armies sought soldiers from traditional "tribal" areas in the northern Gold Coast and Asante, a type military historian David Killingray describes as "untouched by modern ideas of government or commerce, providing non-literate men who would be a clean slate upon which could be written new military codes of discipline and obedience."[89] The British considered the "over-educated" young men from the southern coast as "dangerous and possibly left-wing."[90] Paradoxically, European officers in colonial armies attempted to impose their own doctrines governing hygiene, gender roles, and nuclear families within larger projects of "detribalizing" Africans through military discipline.[91] Like other administrative structures, colonial armies rarely had the resources or will to follow through on their plans to reorder indigenes' social lives, however, so campaigns to regulate soldiers and their families' bodies rarely turned out as planned.[92]

After independence, the remnant of British officers and Ghana's own military and political leaders integrated a version of modernity into the Ghana Military Academy charter that more closely reflected the values of nationalist intelligentsias that bridged the colonial and postcolonial periods, including an emphasis on literacy, nationalism, and bureaucratic efficiency.[93] The modern officer, the charter declared, had "a sound education in appropriate academic and military subjects [and] a wide interest in the current problems of world affairs."[94] The eighteen-month curriculum consisted of three academic terms, focusing on military history, mathematics, science, and French. A fourth topic, "Modern Subjects," was the only academic subject the charter recommended be included in each of the three academic terms.

The aims of "Modern Subjects" were: "to promote and encourage the study of the Commonwealth and its relationship with the rest of the world in the 20th century"; "to develop the ability of the officer cadet to deal with ideas and to form considered judgment as part of a sound liberal education"; "to create and sustain his interest in human affairs and to help him to be aware of contemporary problems"; and to "develop his powers of self-expression both verbally and in writing."[95] "Modern Subjects" would require cadets to develop "historical perspective, and the habit of clear thinking," since the "ability to express thought in simple and direct language and on paper are to be regarded as of great importance than the pure academic knowledge."[96]

What the British officers who prepared the charter assumed made Ghanaians into "modern subjects" closely matched the categories of psychological transformation US modernization theorists also articulated. This is not surprising, since ideologies of development and modernization had flowed easily across the Atlantic for a century.[97] Nationalism, faithfulness to the Commonwealth, literacy in English, opinion-forming, and "clear thinking" resemble the "preconditions" Rostow and Lerner identified as essential for "traditional societies" to abandon.[98] British officers' prescriptions for insulating former colonies (or those Britain was considering granting independence) from communist insurgency matched US strategies.[99] Communism appealed to the "rootless" and "impatient" in colonies, Lucian Pye argued in *Guerrilla Communism in Malaya*, when efficient political and administrative forms failed to fill the vacuum left by weakened traditional social ties.[100]

"Modern subjects" were a bundle of Western ones, and filling the gap in Ghanaian students' understanding of modernity apparently required as much effort as correcting their deficiencies in mathematics, English, and science combined.[101] US Ambassador Russell was pleased with himself that day in 1961 for donating books that offered Ghanaian cadets extra instruction in "modern civilization."[102] Including "Modern Subjects" in the charter of Ghana Military Academy reveals some of the contest over who had the authority to define "modernity," to assign legitimacy to knowledge, and to identify which characteristics of postcolonial nationhood were essential.[103] Modern knowledge, Benedict Anderson suggests, furnished moral importance even to colonized populations, so even externally derived military training might still serve nationalist ends by reinforcing the modernizing elite's authority—either in the person of Nkrumah or the colonels who later overthrew him.[104] To the extent that "Modern Subjects" looked like a template for creating a reliable postcolonial elite comfortable and conversant with former colonial structures, it portended the eventual struggle between the military and the fiercely anticolonial Nkrumah over which versions of modernity and nationalism would survive in Ghana.

When he sacked Major General Henry Templar Alexander, the last British officer to serve as Ghana's Chief of Defense Staff, in September, 1961, Nkrumah symbolically abandoned the steady, incremental improvement of Ghana's military along the British model, but he did not abandon international sources of military training.[105] "Ghanaianization" eliminated British officers but retained the Canadian training staff. To Nkrumah, Ghanaianization was equally about creating a global or pan-African as a Ghanaian image; retaining British chiefs of defense staff before 1961 had been a case of choosing Ghana's national interests—specifically, institutional coherence with the British model—over pan-African interests. When Nkrumah eventually sent seventy-five Ghanaian officer cadets to the Soviet Union for commissioning training in 1961, he was also attempting to balance a global, nonaligned agenda with his own domestic one. Alexander had strongly opposed Nkrumah's plan to employ Soviet training on practical grounds, arguing that incompatible models of training, equipment, and operations would result. Alexander had said that same thing about the mixed Israeli and Indian program to train Ghanaian air force pilots in 1960, ultimately dismissing the Israelis and retaining the Indians because they followed a Commonwealth model.[106] And while Nkrumah publicly aligned himself with the Soviet bloc in a number of ways, Ghana continued to accept military training and military equipment from NATO as well as Soviet sources.

Both Nkrumah and the British-sympathizing senior Ghanaian officers who eventually overthrew him explicitly accepted that any method of recruiting, training, and promoting officers in Ghana's military would be an international effort. Foreign military academy staff produced syllabi for Ghanaian use; foreign officers taught at Ghana Military Academy until Ghanaians could be found; Ghanaian cadets would travel to any service academy that would accept them; and, beginning with the first class of twenty-nine Ugandan cadets to graduate from Ghana Military Academy in 1965, Ghana's own education and training apparatus would serve other African states as well.[107]

In Ghana, international education served competing uses, both of which were tangential to building up the state, yet both upheld "loyalty" as a primary value. British-trained officers leaned heavily toward their martial traditions in the Commonwealth, while Nkrumah sought to balance the lingering Commonwealth and NATO influence with Soviet-oriented training. Despite dire warnings and grave consternation at the time (which may have further encouraged the British-trained officers to overthrow Nkrumah in 1966), the eighty-eight Soviet-trained officers then in the Ghana Armed Forces were equally competent and earned equal promotions.[108] The source of training—East or West—did not matter considerably. Meanwhile, Nkrumah admonished the Ghana Military Academy's 1962 graduating class that the foremost quality required for an officer was loyalty. "Loyalty must transcend all personal interests and ambitions, tribal, or regional considerations," he told them.[109] "Loyalty demands of you that you place the interests of the State above all others, and all your actions must be guided solely by that consideration." This may have simply been an attempt to encourage the officers to inform on one another, as he continued: "You must beware of any covert attempts to undermine your supreme loyalty to the State and you should bring them to the notice of the authorities."[110] But Nkrumah's address, like the one he gave the following year, clearly privileged the state, asserted the military's duty to the state, and, although it mentioned the party, did not demand allegiance to either it or to himself.[111]

Meanwhile, a Canadian officer in Ghana in the early 1960s recalled the "wrenching experience" of observing army officers and cadets as Nkrumah led the country from parliamentary democracy toward a one-party state. "We at the military academy continued to give lectures about the Army's responsibility to be loyal to the duly elected civilian government," wrote former instructor and Canadian Lieutenant Colonel G. D. Hunt, "to which the cadets listened respectfully but, I'm sure, with inner amusement at this white man's fairy tale."[112]

What makes Hunt's recollection of the "white man's fairy tale" of army loyalty to the civilian government so poignant is that he himself—a Canadian officer before a class of Ghanaian cadets—was part of the mirage. Different aspects of fantasy adhere to the concept of military loyalty. Civilian control of the military was, at best, a Western illusion, if not an imposition.[113] In other places in West Africa, postcolonial nationalism successfully enveloped previously strong corporate bonds between companions, and there was some reason to hope that nationalism would subsume military leaders' ambitions in Ghana.[114] A decade later, scholars and soldiers tried to correlate widespread and persistent military rule in Africa with precolonial governing traditions, but the contemporary influence of Western social science required extranational reference points to explain most aspects of Ghana's military evolution.[115] So while some Ghanaian officers (and their Canadian instructors) may have desired Ghana's civil-military relationship with the Ghanaian state to resemble a Canadian or American-style model, that model was an incompletely realized ideal even in the West.

* * *

The new Ghanaian military was not simply one social movement assuming its place in the nation. Nkrumah's cautious admonition about loyalty to the state and Canadian and Soviet commissioning training for Ghanaian cadets were partly examples of the triumph of the modern nation-state. But the political tragedies each attempted to head off (Nkrumah's "covert attempts" and eventual coup; Hunt's "wrenching experience" of military acquiescence to one-party rule) affirm the presence of a political ideal of military noninterference in politics that all participants in international military training acknowledged, even if such training only weakly transferred the ability to attain such an ideal. If both sides saw in their version of the modern military the closest thing to pure nationalism, they arrived at that assumption via international routes. Ghanaian military personnel, then, were adept in national and international cultures, not to the neglect of either. Recognition of a similar "double-cultured" identity occurs more often in the historiography of the African diaspora—which offers another useful way to understand how individuals related with international military travel—but it applies in this discussion of Ghana's domestic military institutions as well.[116] On the other hand, the "double-cultured" identity also implies the ability to "come and go freely between rustic and urban circles."[117] Ghanaian military personnel seemed conversant—even fluent—in international military standards for political behavior with a much more limited ability to enact that identity in practice in governance. In this case, the evidence suggests that, for individuals, international military standards could weaken national bonds as well as strengthen them.

## American Assistance and Ghanaian Internationalism after 1966

The evolution of American military assistance policy to Ghana following the 1966 coup illustrates a new combination of pragmatism and idealism in Americans' understanding of their role in Africa. Ghana's military leaders assured the United States that their military incursion was justified and would be brief, but President Johnson and his advisors still considered military aid to Ghana a multinational effort and resisted expanding American assistance. Meanwhile, the Vietnam War sapped domestic support for American military assistance in general. The United States' ongoing international military education and training program with Ghana exemplified how, despite some glaringly contradictory results, foreign policy makers placed even more faith in the long-term benefits of training foreign military personnel in the United States. The United States still lacked concrete evidence of the programs' effectiveness, but as the academic consensus that had initially supported military modernization crumbled in the late 1960s, the programs remained intact with new ideological scaffolding.

When Ghanaian military and police officers took over the state in 1966, they drastically curbed Ghana's cooperation with communist states, but Ghana's significance as a pan-African leader was already shrinking. Their action represented a local response to specific political and economic conditions in the Ghanaian armed forces and in Ghana's balance of power.[118] On the other hand, between 1966 and 1969, when the National Liberation Council restored constitutional rule, the regime consciously expressed its national agenda according to international norms. From justifying the initial coup and military rule in general, to defending Ghana's defense policy and reliance on international training programs, Ghana's military leaders altered, but never abandoned, the international reference points for the national military. While they continued to believe in the military's power to modernize Ghanaian society, they retained external models of modernity that employed broadly internationalist rhetoric for support.

* * *

Few in the United States or Ghana were surprised when British-trained military and police officers overthrew President Kwame Nkrumah in February 1966. Nkrumah had escaped assassination or overthrow at least seven times between 1962 and 1966.[119] While there is no public evidence the United States explicitly participated in the successful coup, it maintained close contact with the men the Americans knew were preparing the coup. The State Department had waged "psychological warfare" to undermine Nkrumah's domestic support in Ghana since 1964, for example.[120] President Johnson considered the event a "fortuitous windfall" and sent congratulatory messages to the National Liberation Council, but offered only token support for the new regime.[121]

Ghana's new head of state, Lieutenant General J. A. Ankrah, made little headway convincing the United States to expand its economic and military assistance after the coup. He knew he had US political support and delivered captured Soviet anti-aircraft guns to the United States as a gesture of goodwill, but he wanted economic assistance in the form of an International Monetary Fund "line of credit," considerable agricultural aid to avoid a famine "within the next few months," and an anti-aircraft radar and missile system of his own to defend against external threats.[122] The State Department provided $35 million in economic aid but feared "Ghanaians have exaggerated expectations of what the US can do for them."[123] Officials at the State Department knew that Ankrah was disappointed with the "relative paucity of support" that other Western countries—especially Britain—had provided, but they resolved only "to press the Ghanaians toward the U.K., as its traditional supporter."[124] Congress would not support expanding military assistance beyond its current, limited program, so Secretary of State Rusk directed the embassy in London to press the British and Canadians to reassess their ability to help Ghana.[125] Ankrah exchanged warm letters with President Johnson in the spring of 1966, culminating with a visit to Washington in October 1967.[126] Johnson told Ankrah that the United States would "continue to train Ghanaian officers in the US, but we hope Ghana will look to the UK and Canada for the bulk of its military needs."[127] Significantly, the United States still considered training Ghanaian military elites part of its strategy—perhaps its *only* strategy for military assistance to Ghana.

Ghana's military leaders seemed to have convinced American policy makers that they had acted judiciously and were working to restore constitutional rule as soon as possible. Within a month of the coup, Ankrah sent a warm but slightly desperate letter to President Johnson describing his rationale for seizing power and intention to restore democracy as soon as possible.[128] In his reply, Johnson acknowledged the coup's necessity, offered "a few thousand tons of surplus wheat or rice," and held out hope for a return to democratic rule.[129] Coup leader Colonel Akwasi Afrifa credited the army's commitment to democracy to traditions forged at the Royal Military Academy, Sandhurst.[130] By spring, 1967, William F. Gutteridge of the Royal African Society published an article in *African Affairs* that recognized the persistent psychological importance of Western military training on Ghana's military leaders but also drew a distinction between imagined Western military traditions of "non-intervention in politics" and the reality behind military influence over Western politics.[131] Gutteridge argued that, rhetoric aside, Western militaries regularly intervened in their states' politics. Meanwhile, Ghana's coup leaders seemed genuinely distressed about their political role. Further, although foreign aid for African militaries, particularly the training of officers overseas, may provide them some encouragement to intervene in politics, to Gutteridge, the purpose of their intervention was more likely "determined by particular local conditions than by the influence of the source of assistance."[132] Gutteridge gave the Ghanaian military the benefit of the doubt in 1967 and provided some academic cover for continued American programs to train Ghanaians. In June 1967, Ankrah again assured Johnson that he had begun "steps toward civilian rule and the re-establishment of institutions essential to democratic Government."[133] Prior to Ankrah's October 1967 visit, Rostow reassured Johnson that "Ankrah and his colleagues haven't behaved like the usual military dictatorship," and he praised Ankrah's National Redemption Council's steps to draft a new constitution and prepare for elections.[134]

Despite their relief at Nkrumah's overthrow, American diplomats did not wholeheartedly support Ghana's coup leaders. For one, they suspected Ghana was requesting weapons it could not afford and did not need, given its insignificant external threat. While fending off Ankrah's requests for anti-aircraft radar, Secretary of State Rusk advised the American ambassador in Ghana to avoid discussing sales or grants of military equipment and "do nothing to encourage" the Ghanaian government "on prospects for US military assistance" of any kind.[135] The costs of providing "adequate defenses against vague threats" would be too high. Further, the State Department considered the army itself a threat to Ghana's stability.[136] At 15,000 men, the recently expanded army was the largest in West Africa, served little useful purpose, and consumed far more national resources than Ghana could afford. But given Ghana's high unemployment, it could not be reduced without serious danger.[137] Franklin H. Williams, the American ambassador in Accra, feared that the failed counter-coup of April 1967 had shown the potential for more political violence. "*The army is too big.* The army is larger than the country needs or can afford," he reported to Washington.[138] Yet even those army officers who were "more interested in quality than quantity" recognized that it was "not an opportune time to cut back" and risk a counter-coup or add to Ghana's unemployment.[139]

By 1967, the president and the secretary of state also knew that domestic and congressional support for expanding military assistance to Ghana or any other African country had withered. Congress had cut the Military Assistance Program by nearly half in 1967 and expected to cut it further in 1968.[140] In May, Rusk told the US ambassador in Ghana that "Congressional distaste" would prohibit any new grant-based military assistance programs.[141] "It [is] not a question of being opposed to aid or help to individual countries," Rusk cabled. "He personally believed" that the United States spent too little on aid, but with the large federal deficit and "heavy Viet-Nam budget," he expected Congress to slash all foreign aid further.[142] Under Secretary Katzenbach advised President Johnson to give Ankrah the same news in October, during the head of state's visit to Washington. "We value Ghana's constructive influence in Africa, and we want to let [Ankrah] know it," Katzenbach told Johnson. "*At the same time, we want him to understand that there are limits to what we can do for Ghana.*"[143] "There is great hostility in the Congress to military aid," he wrote.[144] "We will continue to train Ghanaian officers in the US," he added, "but we hope Ghana will look to the UK and Canada for the bulk of its military needs."[145]

As the president, State Department, and Defense Department all planned for Congress to cut drastically all future foreign military assistance programs in 1967, they reiterated their support for programs that brought individuals from the Global South to the United States for military training. In his detailed report to the president describing the expected impact of congressional cuts on the Military Assistance Program, Defense Secretary McNamara considered the United States' specific military objectives in Africa "quite limited."[146] Except for minor programs to continue to lease military facilities in Ethiopia and Libya and offset Soviet influence in Algeria, the United States predominantly aimed to "improve the capabilities of the armed forces to maintain internal security, where appropriate, as a basis for orderly national development."[147] McNamara claimed that of all military assistance, there was no substitute for training programs in the United States. "Such training is an effective augmentation of our diplomacy," he argued, "and should be continued."[148] International education and training absorbed 14 percent of the $375 million Military Assistance Program in 1969 (itself down from $1.6 billion in 1962). On the other hand, despite the 1963 projection to spend $150,000 on Ghanaian training grants, in 1967, the United States spent only $25,000 for twelve Ghanaian students' training in the country. Thirteen arrived in 1968; fourteen in 1969 as the United States began sponsoring Ghanaian officers for Fort Leavenworth's Command and General Staff College. The main justification for military education and training programs had shifted from being modernization-based toward a more pragmatic supplement to diplomacy as the political environment in Africa shifted toward military governments, but faith in the programs' power to transform society continued in both the United States and Ghana.

* * *

Ghana's new military leaders left behind clues to help understand this shift. In February 1967, the director of public relations for the Ghana Armed Forces published the first volume of *Ghana Armed Forces Magazine*. This journal, head of state Lieutenant General Ankrah stated in an editorial in the first issue, would inform the armed forces and police service of "the affairs of our country," would "provide a forum for the sharing of experiences," and would "serve as a chain of brotherhood among officers and men."[149] The first several issues included local interest stories, updates from different branches of the service, and news of armed forces sporting clubs. The magazine, which renamed itself *Ghana Armed Forces News* after 1969, was always a mouthpiece for the current regime, military or civilian, but in attempting a wide coverage of political, military, and social events in Ghana, it published articles from a variety of sources, including chaplains, enlisted correspondents, editorials from Ghanaian units outside Accra, and occasionally letters to the editor.[150] Anonymously penned articles celebrating the coup, with titles like: "The Coup: A Ray of Hope At Last," "The Expected Day!" and "The Rationale Behind Military Intervention" filled several pages as well. Some of the magazine's more thoughtful articles reveal a subtler approach to the coup and to the magazine's purpose explaining it. They also reveal a complex dialogue between Ghana's armed forces, Ghana's government, and international standards for military professionalism, diplomacy, and journalism.

In the first article of the first issue of the *Ghana Armed Forces Magazine*, Ankrah offered the standard justification for the coup, but he did so in a surprisingly multicultural way. After his explanation, and before declaring the magazine's purpose, Ankrah included an elegy for the coup's relatively few casualties—seven soldiers and four civilians dead—which he merged with Abraham Lincoln's Gettysburg Address. "We do well to remember those of our comrades who paid the supreme sacrifice by laying down their lives in selfless devotion," he wrote.[151] "We who are left behind are morally bound to dedicate our lives to the democratic ideals for which they gave their all and their best." He continued: "Let this be our daily resolve that these our dead comrades shall not have died in vain and say with that great champion for civil liberties that this nation, under God, shall have a new birth of freedom and that we in Ghana shall thrive in democracy, freedom, tolerance, goodwill, prosperity and peace."[152]

Ankrah was not writing strictly for a foreign audience here, although early editions of *Ghana Armed Forces Magazine* appeared in libraries around the world.[153] His next sentence, "This new journal is earnestly commended to your patronage," spoke directly to the men of the armed forces and police service. Nor is it likely that he plagiarized Lincoln's stirring prose expecting nobody to notice—he specifically cited "that great champion for civil liberties." Instead, in the first article of volume 1, issue 1 of *Ghana Armed Forces Magazine*, the commander of the Ghana Armed Forces, through the cooperation of the director of public relations, described the political and moral role of the military in Ghanaian society in terms that specifically merged Ghanaian and international language, symbolizing the Ghanaian army's simultaneous presence in both *milieus*.

Through the magazine, Ghana's military leaders regularly blended national and international rhetoric to explain domestic politics and Ghana's relationships with its regional and global partners. In 1969, with Ghana on the verge of returning to constitutional rule, *Ghana Armed Forces News*' "The Rationale Behind Military Intervention in Politics" integrated Feliks Gross's *The Seizure of Political Power in a Century of Revolutions*, Finer's *The Man On Horseback*, and Huntington's *Changing Patterns of Military Politics* in order "to discuss the conditions under which coups and Revolutions can be deemed inevitable" and partially excuse the Ghanaian case.[154] In articulating Ghana's defense policy, Colonel A. K. Kattah compared Nkrumah's "muddled up" priorities and "expensive white elephants," which failed to follow "age-old strategic and diplomatic tenets," against American, British, and Israeli strategic concepts, which "were clear and precise and were based on their fundamental conception of what entailed dangers to their national interests."[155] In 1968, public affairs officer Captain S. G. Amoo began his article describing the role of the director of public relations by quoting James Forrestal, the United States' first Secretary of Defense, without irony: "I know of no task that is more complex, except possibly the task of government itself, than that of engendering in a democracy an appreciation of the role of the Armed Forces."[156]

Raising the Ghanaian public's appreciation for the role of the armed forces also merged local and international discourses. Early editions of *Ghana Armed Forces Magazine* attempted to remove the stigma of military service and rehabilitate the "Abongo."[157] Amoo's "Selling the Forces to the Taxpayer" imagined "these strong Abongo boys" coming to Ghana's economic aid, mobilizing military men and equipment on behalf of Ghana's cocoa industry.[158] When they did, in December 1968, hauling 1,400 tons of cocoa in army vehicles from the interior to the coast over badly degraded roads, the *Ghana Armed Forces Magazine* celebrated the new rapport with locals in which "any fear or mistrust which these inhabitants might have entertained about the military gave way to harmonious relationship between them."[159] In volume 1, issue 1, public affairs officer Lieutenant A. A. Enninful attempted to put the Abongo stereotype to rest. For Ghanaian parents, "the lofty ambition for their children was to give them a high education for a white-collar executive position, a doctor, or engineer—but not a soldier."[160] Enninful imagined a past scene in the home of a young man announcing his plan to join the service to his parents in which "the wailing of mothers" usually accompanied "the least inclination of their sons to join the army": "'Abongo? Not for all the breath that holds me—not in my lifetime. You can join *your abongo* when I'm dead and gone you damned one; and even that my ghost shall keep haunting you.'"[161]

Now, Enninful wrote, the well-organized and disciplined armed forces have replaced the traditional *Asafos* of old, and "most parents are beginning to realize the Armed Forces is a useful arm of the nation, with a large fund of bright prospects for the youth."[162] The armed forces liberated Ghana, Enninful suggested, when it "took up arms one morning to chase out Kwame Nkrumah, his myrmidons, and his Corrupt People's Party organization from our midst."[163] As Ghana's economy struggled, the armed forces offered an alternative, often international, path for those who could not afford to educate their children. "A number of young officers are currently on study leave in Ghanaian universities," he wrote. Even more, "Elsewhere in European universities, there are lots more studying engineering or doing post-graduate courses."[164] Even as an avenue for advancement within Ghana and as an antidote to the stereotype of the parochial, uneducated Abongo, the armed forces advertised itself through the international opportunities it offered. As Chapter 3 explores more, many Ghanaian soldiers recalled that the opportunity to travel drew them to the service.

International education and training provided multiple fora for Ghanaians to articulate national objectives. When a group of senior American military officers from the National War College visited Ghana in spring 1968, the Ghanaian chief of defense staff, Air Marshall Michael Otu addressed them at length about Ghana's defense policy. The Americans, Otu suspected, "may be dying to hear from our lips, our own version of the defense or justification for the forcible overthrow of the old regime," but he defiantly refused to give it.[165] Instead, he summarized Ghana's domestic and regional security roles and called the West to task for its failure to ensure justice in Rhodesia and South Africa. "There is no doubt that the sincerity and motives of Western powers towards Africa and African problems is facing a severe test," he told them.[166] "If only they would not betray the humanistic ideals upon which their countries and their greatnesses have been founded the solution may not be too far to seek."[167] On the other hand, as Ghana's armed forces sought greater efficiency and expertise, he recognized the ongoing importance that "both officers and men take advantage of higher military courses overseas not available in Ghana."[168] "Here," he continued, "I must express my deep gratitude to countries like yours, Canada, and UK which readily provide us with these training facilities."[169] Finally, he addressed Ghana's industrialization and economic development, the National Liberation Council's path toward restoring constitutional rule, and the armed forces' potential as a resource "for the development of the country in various fields, including: building of bridges, roads and townships; conveying essential supplies and foodstuffs to afflicted areas during emergencies and national disasters; [and] medical evacuation by air."[170] Ghana's chief of defense staff in 1968 clearly considered the military the most modern institution in Ghanaian society.

Modernization theory declined in the West in the late 1960s, but Ghanaian voices still reaffirmed old assumptions about the military as the most organized, cohesive, or rationalized institution in Ghanaian society.

When the civil state crumbled during years of military rule, those assumptions became self-fulfilling.[171] From its very first issue, *Ghana Armed Forces Magazine* repeated the claims, although not without contradiction.[172] "As a large complex organization involves almost every aspect of the national, political, social and economic life, the Armed Forces should, without any reservation, maintain leaders to transcend . . . elementary military qualities," it argued in 1969.[173] Military leaders worked "in close harmony with civilian diplomats, scientists, economists, legislators, industrialists and experts in other innumerable fields," it continued. Therefore, "they should have the ability to make sound judgments on matters relating to a large percentage of the population of the country."[174] As if recognizing the likelihood that military rule would return to plague Ghana, the final issue before the first round of elections in 1969 both praised and cautioned military personnel. "The Armed Forces is the most organized institution in the country," read an unattributed article titled "Military Personnel and Politics."[175] "It is a closely knit body, and the nature of our job requires absolute solidarity and loyalty to our superiors, inferiors, and one another." While admitting Ghana's "first attempt at self-rule was a pathetic failure, and should be better forgotten but for the fact that we need past experience to guide future actions," this article still warned, "it is also in our interest to eschew politics."[176]

Even when a military junta that vocally proclaimed its preference for American assistance took power in Ghana in 1966, the United States did not expand its military assistance program beyond its ongoing training and education mission. This was not simply a case of the United States preferring the new Ghanaian government as a Cold War proxy. American diplomats in Accra and Washington suspected the military regime's capacity to govern and refused to fulfill the military government's other requests for weapons and equipment. Sponsoring training and education in the United States, on the other hand, easily accommodated military rule because it promised to deliver a different kind of result. Especially when strong congressional reaction against failed American modernization-turned-military missions in Asia and Latin American threatened foreign assistance budgets, both the State and Defense Departments reiterated the usefulness of education and training programs. The Americans' response to persistent Ghanaian requests for more military aid exemplified how the State and Defense Departments responded to congressional pressure to downsize all security assistance programs: Military education and training programs were the best way to provide corporate benefits to the recipient nations and Cold War allegiance—at low cost—in the long term, but they did so at a personal, psychological level.

Military rule in Ghana, therefore, did not require the United States to reconsider its assumptions.

* * *

As they had since independence, members of the Ghana Armed Forces after 1966 mingled local and international language to articulate their national agenda. When validating the coup or describing military rule, they carefully selected rhetoric that merged fluency in internationally recognized discourse—like the Gettysburg Address—with local idioms—like the Abongo stereotype—to reveal a transnational understanding of the role of the military within Ghanaian society. Between the first and second coups, even the most stalwart defenders of Nkrumah's overthrow carefully acknowledged internationally recognized norms of civilian control of society. On the other hand, by recycling previously fashionable academic rhetoric affirming the military's potential as the most modern and organized social institution, they kept alive and modified an international intellectual import that had largely gone out of fashion in the West. This combination of local and international references portended ongoing military involvement in Ghanaian government.

## Ghana's Return to Military Rule and the Limits of American Military Education

Scholars did not universally condemn the first Ghana coup in 1966, and Ghana's return to civilian rule in 1969 suggested the initial foray into military rule might have been an aberration or a correction. In July 1970, as part of the Pugwash movement, the Adlai Stevenson Institute sponsored the first Pan-African Regional Symposium on Disarmament and Development at the University of Ghana. The symposium drew tame conclusions about civil-military relationships in Africa. It acknowledged that excessive defense budgets in Africa diverted essential resources away from economic and political growth and made African armies more likely to intervene in politics; but, it argued, with proper guidance armies could "be directed to nonmilitary objectives" and "make a tremendous contribution to physical development."[177]

The conference's choice of venue—the University of Ghana in 1970—is more telling. Ghana was the first African state to return to civilian rule after a coup, and so it inspired academic speculation cloaked in hope that other military-run states might also return to constitutional rule. When Ghana's education minister addressed conference attendees, he argued that small states and new states depended even more heavily on the international order, especially the United Nations, not just for their own security, but also as the most effective path to national greatness.[178] Transnational cultural and educational exchanges directed toward industrialization and economic integration offered the best alternative to militarization, Minister William Ofori-Atta argued, asserting that international cooperation was the essential component for avoiding economically debilitating military competition.[179] For Ofori-Atta and the conference's other scholars who feared that Ghana was still particularly vulnerable to relapse into military rule, military intervention would be more likely to produce political decay than political development continent-wide, but international intervention in education and training was the antidote to military intervention in politics.[180]

Ghana's initial popularity as a proto-independent African state, its descent, revival, and re-descent into military rule, made it a common example for scholars of military rule in Africa and a representative case study for American military assistance policy over time. Ghana's second coup in 1972 and the persistence of military rule elsewhere in Africa required scholars to reassess military rule's justice and prospects for success. Usually, observers condemned the military leaders who, after overthrowing civilian governments, were unable to govern more efficiently and unwilling to return to constitutional rule. Samuel Huntington had argued in 1968 that "military explanations do not explain military interventions" and that coups were simply one manifestation of the general polarization and politicization of social and political institutions in underdeveloped societies.[181] After Ghana's second coup, however, most scholars considered the military itself the problem. The first coup violated previously "inviolable territory"; the military's appetite "had grown perhaps by feeding"; and "the aftermath of military intervention is military intervention."[182]

One reason American policy makers had such difficulty evaluating the long-term effects of their assistance programs was that recipient nations had unique, and often changing, uses for American aid. The case of Ghana demonstrates that even in a state that struggled to balance power between military and civilian political factions, the United States' military exchange program assumed disproportionately large significance only in the American perspective.

* * *

Kofi A. Busia, Ghana's prime minister between 1969 and 1972, recognized that domestic political and economic conditions had distinctly international causes but that international military assistance rarely emerged as a solution. The ruling military council had established Busia's Progress Party into power in 1969, and even though Busia rejected Nkrumahism, his vision of Ghanaian nationalism was no less pan-African. In October 1969, he told a gathering of American journalists at the National Press Club:

> We shall not be tied to the apron strings of any country, whether East or West. We shall make our own decisions and take our stand on all international issues on the basis of two considerations: the interests of our country within the context of our international obligations; second, our consideration for the welfare and peace of the world on which our own progress and prosperity will ultimately depend.[183]

The extent to which Busia or future leaders were willing to condemn American policy toward South Africa offered one measure of their diplomatic allegiance to the United States. Busia trod a precarious route between appearing as an unpopular "dialogue man" to ordinary Ghanaians and placating the Americans in order to secure economic assistance. Ghana's urgent need for debt relief or restructuring convinced Busia to moderate his critique of American policies, especially its toleration of apartheid in southern Africa.[184] Like other leaders across the Organization for African Unity (OAU), Busia struggled to keep alive enough diplomatic goodwill with the United States to sustain the economic assistance he needed when Ghanaians overwhelmingly opposed both apartheid and the United States' role in preserving it.[185] At home, Busia criticized apartheid, even encouraging the OAU to do more to train anti-apartheid "freedom fighters," but sent mixed signals to the world by siding with President Nixon in the United Nations to encourage "dialogue" over sanctions.[186] Meanwhile, he pressed the United States for more economic assistance. Ghanaians may be grateful for US aid, he said, but "as a democratic country which underwrote Nkrumah's tyranny . . . they should do more to underwrite the system which we are trying to build here."[187] Busia—the civilian statesman—used his limited diplomatic leverage to both prod and praise American foreign policy in Africa. Later, shortly after Colonel Acheampong overthrew Busia's government, the American-trained military officer rejected that cautious path, reaffirming Nkrumahist priorities of "positive neutralism" and "nonalignment," and announcing that Ghanaian diplomats would "no longer be advocates of dialogue with South Africa."[188] If acquiescence to other American security initiatives in Africa was one desired outcome of the United States' ongoing military education and training program, Acheampong's repudiation of "dialogue" indicates that objective failed.

The United States repaid Busia's circumspection with ambivalence. In its May 1971 policy planning paper for Ghana, the State Department wished to maintain Ghana's generally favorable outlook toward the United States, but "without trying [sic] ourselves closely to either Busia's government's political designs or its political future."[189] To the Americans, "the Busia government has not come to grips with such problems as endemic corruption, governmental inefficiency and lack of coherent planning."[190] In addition to continuing economic aid—especially food and commodities—the State Department recommended maintaining its current diplomatic "courses of action," including cultural diplomacy programs, "an imaginative and flexible Peace Corps presence," its "information program activities," and academic exchange programs.[191] The State Department also recommended maintaining "at least present funding levels of the small but important [Military Assistance Program] training program." The military might eventually pose a threat to Busia's government, but the State Department's country team thought the military leadership closely aligned with Busia's Progress Party.[192] On the other hand, since the military remained "the major force capable of unseating any government," State Department staff feared Ghana's weak "military and government control mechanisms" made it vulnerable to a "captain's coup"—overthrow by junior military officers.[193]

Military assistance formed only a small part of American foreign policy with Ghana before the second coup in 1972, but the United States' ambivalent reaction to the coup belies the emphasis it placed on keeping that small military education and training program operating. In their draft policy planning paper, composed six months prior by US Embassy staff in Ghana, the country team included another sentence in the recommended "course of action" regarding the Military Assistance Program cited above. It read: "Maintain at least the present funding levels of the small but important MAP Training Program. *This is the only program we have directed toward the politically crucial military establishment.*"[194] That second sentence disappeared from the final, state-approved document in May 1971.

The embassy's country team prepared their draft in December 1970, and, in retrospect, that statement seems prophetic. Just eighteen months earlier, Lieutenant Colonel Ignatius Kutu Acheampong had completed his year at Fort Leavenworth's Command and General Staff College in the United States. He seemed to be precisely the kind of prominent military establishment figure the Americans intended to inspire. One year after this draft, however, he seized control of Ghana's government while Busia traveled in England. Yet to the extent that the Americans noticed the connection between their International Military Education and Training Program and

Ghana's new military government, it barely affected the United States' relationship with Acheampong.

In the first weeks after the coup on January 13, 1972, US Ambassador Fred Hadsel sat tight. Days after the coup, in a cable back to Washington, he mused about the "[the] basic dilemma which has concerned me from [the] inception of [the] coup."[195] Coming to Busia's defense might appear to support democracy in Ghana, but returning to the status quo ante—and the ineffectual Busia—risked "delaying [any] solution to Ghana's fundamental problems."[196] Colonel Acheampong's National Redemption Council (NRC) regime seemed by the twentieth of January to be "apparently in full control of the country," therefore American action supporting Busia might "encourage more extreme governmental action which would be inimical to our interests here and elsewhere in Africa." Hadsel did not assume that Acheampong's rise represented a freshening of Ghanaian-American relations. On the other hand, if the United States should "focus on Ghana's economic problems, putting these above our admiration for Busia and our dismay at a military coup," they might encourage "sensible solutions (if the NRC will take sensible action), [and] seek to achieve a condition in which democracy can be renewed." Hadsel chose the latter path, but not out of confidence that anything about Acheampong's training made him a natural ally with the United States.

Hadsel's first impression of Acheampong and his colleagues on the National Redemption Council was that they appeared "woefully inexperienced in government administration."[197] They seemed united only by a common "disgruntlement over lack of promotions." Still, in the brief biographies of NRC members that he cabled back to Washington and to other diplomatic posts across Africa, Hadsel revealed an unusual attention to the international sources of their military education. Acheampong, Hadsel acknowledged, graduated from the US Command and General Staff College and "has complained about officers his junior being promoted ahead of him and of [former Chief of Defense Staff] Addo's alleged preference for officers trained in [the] UK."[198] Lieutenant Colonel Chemogo D. Benni attended the Indian Military Academy and had graduated from Ft. Leavenworth's Command and General Staff College in 1969. Major Anthony H. Selarmey attended the US Army's Officers' Advanced Armor Course at Ft. Knox, Kentucky, in 1968. Major Kwame Baah earned his commission in India and had served as "staff officer to defense adviser in Ghana's embassies [in] London and Washington."[199] All these officers had received advanced training or been posted abroad, but neither did they attempt to personify "professional" ideals of civil-military separation and Cold War allegiance to the West embedded in American assumptions nor was Ambassador Hadsel surprised that they did not.

In a conversation with the US defense attaché one month after the coup, Acheampong remembered his military training in the United States warmly and "proudly explained that most of the members of the NRC were graduates of US or UK schools."[200] Acheampong "emphasized Ghana's continuing need for education opportunities of this kind" and made special mention of the superiority of American military education, "noting that the preference for UK training imposed on the armed forces by former [Chief of Defense Staff] Addo was part of the past."[201]

Such an endorsement by the new, American-trained leader of a troubled African state ought to have been the ultimate validation of the United States' international military education and training programs, but it was not. It barely registered to American diplomats, scholars, or the Defense Department itself. The American ambassador immediately discounted Acheampong's ability—he was "woefully inexperienced in government administration" and "completely uninformed on international relations and obligations"—and suspected his judgment. Hadsel would recommend sensible solutions, he said, but he doubted the NRC "would take sensible action."[202]

For his part, Acheampong did not demonstrate the emotional attachment or political allegiance with the United States that American justifications of the program touted. He later told Accra's *Daily Times* that he began preparing a military takeover as soon as he returned from the United States.[203] He rarely referred publicly to his American military education, even in the Ghana Armed Forces' official newspaper. By 1977, his official biography made no mention of it, despite describing his other military posts in the 1960s in great detail.[204]

Instead, Acheampong claimed that the he placed Ghana's primary political, economic, and international concerns before his relationship with the United States—diplomatically or militarily. He proclaimed, "My belief in Pan-Africanism is total absolute," including "total commitment to [the] aims and objectives" of the Organization for African Unity, while advocating a mix of continental and national self-reliance.[205] He unilaterally postponed repayment and repudiated Ghana's international debts, arguing, "We are jealous of our sovereignty which we would not like to see mortgaged to any power."[206] He restored diplomatic relations with the People's Republic of China and several Eastern European states, including East Germany, and attempted to revive East German economic development projects Ghana had unilaterally terminated after the 1966 coup.[207] In April 1972, he announced, "The National Redemption Council rejects any form of dialogue with apartheid South Africa, as long as that country sticks to its policy of racism."[208] "The time has come for us to stop trying to be second-rate Europeans," he wrote in

1974. "Our mission is to become first-rate Africans."[209] And like the military government of 1966, he was reluctant to shrink Ghana's large armed forces, arguing that instead of adding to the unemployment rolls, Ghana could harness the army's trained personnel and equipment for a new program of self-reliance called "Operation Feed Yourself."[210]

Ghana's military leaders understood their unpopularity with respect to academic critiques of military rule abroad. *Ghana Armed Forces News* published only one book review in 1976: a critique of Austin's *Politicians and Soldiers in Ghana*, which concentrated mostly on deriding University of Boston historian Valerie Plave Bennett's "Epilogue: Malcontents in Uniform."[211] "It reeks with that 'conservative' American tendency to look at un-American situations with distorted vision," wrote Tom Dorkenoo, editor of *Ghana Armed Forces News*. "The 318 page book has largely failed to give deep insight into the subject or subjects it set out to write about," he continued; "yet it is a book for one to read, if only to know the minds of the authors and what academicians, Ghanaians and foreigners think of the years before and after the advent of the NRC-SMC Government."[212] In short, Acheampong revealed rather closely the same kinds of domestic political and economic concerns as previous and later Ghanaian governments, with an equal understanding of the international nature of their origins. Acheampong's yearlong education at an American staff college made no obvious impression on either his own pronouncements, his willingness to condone American policies toward southern Africa, or on the United States' diplomatic mission to Ghana.

Likewise, the United States' military assistance policy undertook little soul-searching, even after one of its own alumni demonstrated the program's negative potential to produce tragic results. In 1976, Ambassador Shirley Temple Black reaffirmed the long-held American opinion on international military education and training grants to Ghana. Responding to Secretary of State Kissinger's request that all foreign missions evaluate their states' needs for American international military education and training grants (with an eye toward trimming the program), Black assured Kissinger that Ghana depended on American military training. The training reaped significant benefits for American interests, she argued, and professional military education specifically was the most useful for those reasons.[213] Ghana most desired courses "which will enhance professional development," she said, and she assumed that they would give up all other types of training to hold onto their slots for Command and General Staff College and similar advanced education.[214] Those courses were also the "most beneficial to US national interests," she said, because "these are timed for the period during an officer's career just prior to his possible assumption [of] key roles either in the military or in the government."[215] The Ghana government "has planned ahead in its training program on the basis that US assistance could continue for some time," and Black feared that Ghanaians would assume that cuts to the program resulted from "political, rather than inflationary or budgetary" factors, as a form of American "retaliation for foreign policy actions along lines which have been well publicized recently in US press and abroad."[216] Should the Americans cut their military education and training budget, she feared, Ghana might turn to the Soviet Union or Eastern Europe for training courses, which "would not be helpful in any way to US-Ghanaian bilateral relations."[217]

Here, again, an American official projected that Ghanaian participation in American professional military education served the Ghana Armed Forces' corporate needs, secured American interests in Ghana's government, and kept Ghana from seeking military assistance from the Soviet bloc. The Americans even closely managed which American schools they would fund for training grants: Black observed that the Command and General Staff College occurred at a critical juncture in the Ghanaian officer's career—"just prior to his possible assumption of key roles either in the military or the government." She either failed or declined to acknowledge that Acheampong had followed precisely the short path from American staff college to a key role in government—coup leader and head of state. Within two months of this reply to Kissinger, Acheampong's foreign policy so frustrated Black that she temporarily froze all new aid programs to Ghana and recommended her own recall to the United States to reevaluate US-Ghanaian relations.[218] That she still placed such high confidence in international military education and training programs to serve both American and Ghanaian interests suggests that either an imagined or idealized logic underwrote that program. Despite the glaring failure of one American staff college graduate to so transform US-Ghanaian relations according to that ideal, American and Ghanaian policy makers clung to those exchange programs. Even if Acheampong was an outlier rather than the desired outcome of international military education and training programs, the operation and justification for the international military education and training program changed little for the Ghanaian military or for American policy makers after 1972.

## The Ghana Armed Forces Staff College and Military Internationalism for Export

One reason neither the United States nor Ghana overestimated the significance of Acheampong's year at an American staff college was that, before 1976, all senior Ghanaian officers completed staff college abroad.[219] Since Nkrumah's time, the Ghana Armed Forces had struggled to create a Ghanaian staff college. The school's eventual formation in 1976 reveals many of the paradoxes between states' faith in military-based internationalism to transcend national or corporate military identities and the practical difficulty for small states that relied on external (international) assistance programs to produce the equivalent of senior civil servants in a military government. Just a few years after the school opened, Ghana Armed Forces Staff College's "Ghanaianization" plan again reenacted the struggle to develop a specifically Ghanaian—or African, given the number of non-Ghanaian students there—institution. The staff college that resulted negotiated international military norms it had inherited and that continued to be refreshed through ongoing training abroad, and national performance in the context of a military-run government.

* * *

Since independence, no Ghanaian political or military leader questioned the importance of staff college training, but the expense of building a national staff college and the availability of staff college training abroad postponed the Ghana Armed Forces Staff College's creation for over a decade.[220] Kwame Nkrumah founded Ghana's National War College in 1963, intending for it to serve all Africa. By uniting, he reasoned, small African states could manage the high costs of training and equipping their armies and resist being "drawn into making defence pacts with foreign powers which may endanger the security of us all."[221] Lack of funds stymied the National War College plan, but the school Nkrumah founded became the Junior Defense College, which has provided primary staff college training for Ghanaian lieutenants and captains since 1964.[222] After the 1966 coup, the military government accepted more military training assistance from abroad to demonstrate "the country's reorientation to the west" while cutting Ghana's military budget.[223] It tabled the plan to build a Ghanaian staff college. After 1966, Ghanaian officers received staff college training in the United Kingdom, the United States, Australia, Canada, India, Malaysia, Indonesia, and Pakistan, but after 1971, Commonwealth schools required Ghana to purchase tuition rather than receive training slots as grant aid.[224] Before 1972, Busia's government made no effort to fund the staff college plan, but the Acheampong regime revived it in January 1973. Ghana Armed Forces Staff College accepted its first students in January 1977 in Teshie, alongside the Military Academy and Training School.[225]

Avoiding the cost of sending officers abroad was one significant reason Ghana revived the staff college plan in 1973, but the staff college was not strictly an insular national organization.[226] Just as Ghana Military Academy had, Ghana's staff college became a center for pan-African military education. Four Nigerian officers joined the second Ghana Armed Forces Staff College course in 1978, and forty-one Nigerians had graduated by 1992.[227] Starting in 1981, one Tanzanian officer joined each class. Later, officers from Sierra Leone, Uganda, and Zimbabwe attended Ghana's staff college. In 1980, Ghana and Nigeria began exchanging faculty at their respective staff colleges; by 1984, two of each nation's officers taught at the other's staff college.[228]

The staff college's international character revealed Ghana's increased participation in, not segregation from, ideals of global military integration, and it demonstrated Ghanaians' comfort as full participants as both a source and a subject of international military education. Acheampong's 1975 address to the first graduating class of Ghana Military Academy to include a Liberian graduate expressed "the honor of training both officers and cadets from other friendly African countries" and "renew[ed] [Ghana's] offer to all friendly African countries." "Our doors are still wide open," Acheampong stated, "and we shall feel honored to continue to share our experience with them."[229] In the same year, as Ghana struggled to find a site for the staff college and develop a curriculum, Ghana's chief of defense staff, Lieutenant General Lawrence Okai, asked the Canadian and British armed forces to provide faculty and staff.[230] More than financial reasons forced Ghana to seek British and Canadian help. Okai graduated from the British staff college at Camberly in 1966 and intended for Ghana's school to meet "Commonwealth and international standards . . . as this would enable the Directing Staff [the faculty] to be exchanged between Ghana and other Colleges of Commonwealth countries."[231] It also equipped Ghanaian graduates to lead international contingents of UN soldiers in peacekeeping missions abroad, which Ghana had begun again in earnest in 1974.[232] From the beginning, the Ghana Armed Forces intended for its staff and graduates to participate in staff assignments and exchanges with other nations, assuming equality even with the British and Canadian schools that sent faculty to Ghana.

Just as Ghana's domestic economic policy of "flexible self-reliance" merged national and international influences, financing, and participation, Ghana Armed Forces Staff College exemplified the dynamic flow of local and global in military education that forms one aspect of Military Internationalism. The campaign for "Ghanaianization" in the late 1980s reveals some of this dynamism. One former commandant recalled that the "initiative to Ghanaianize the Staff College completely" began in response to the "heavy" Canadian and British influence over the college and its curriculum.[233] Ghanaianization was about more than just finishing the decades-long process of replacing British influence from Ghana's armed forces, but that was an important first step. At the senior staff college's founding in 1976, a seven-person Commonwealth Military Advisory team, headed by a British colonel who also served as the deputy commandant, augmented the small Ghanaian faculty.[234] During the initial negotiations in 1975, Britain requested that a British officer should fill the deputy commandant's position in order "to prevent the British officers from coming under direct command of Ghanaian authorities."[235] Over the next decade, British influence declined but never disappeared. By 1987, the Ghanaian faculty had grown to eleven and the British contingent had shrunk to two—but a British colonel still filled the deputy commandant position.[236]

At least one former staff college commandant hated that structure. The Ghanaian commandant was the unit commander; in his absence, the deputy commandant performed the commandant's duties, including commanding Ghanaian officers, but he was technically not subject to Ghanaian laws. "It was not proper," the former commandant insisted. "The deputy therefore was performing the Commandant's duties, so he was performing Ghanaian duties [but] he [was] not subject to Ghanaian law."[237] "When we did the review," he recalled, "we maintained him, but this time not as the deputy commander, but as the head of studies—[an] academic [position]."

"The review" occurred in 1991, when the commandant, Brigadier I. G. M. Kpeto, closed the senior division of Ghana Armed Forces Staff College for one year to rewrite its syllabus, reorganize its faculty, and expand its facilities.[238] In the 1980s, a former commandant recalled, "It was a very strong challenge. . . . We were teaching our Ghanaian armed services as distinct from both America, Britain, and Canada. These were the places where we were attending the staff colleges, of course, also India. We were attending Indian Staff College, Pakistani Staff College. So all these things, we have people training here, people training there. Which personality are we going to teach? We [had] to reach a common one."[239] To stress Ghana's difference, or at least its independence, Kpeto directed the Ghana Armed Forces Staff College's one year sequestration. A staff officer recalled, "We wanted to see a lot more of Ghana asserting itself. Its own strategy, its own policies, not copying American strategy, or Britain. They are good for learning, but if the Ghana Armed Forces had to go to war, we would fight here, not outside, so let's try and know our own territories, and use our own technology to develop our military."[240] Ghanaianization did not signal an end to the prior faith, partly borne out of Ghana's inequality with larger Western militaries, that all professional education was equally valid, regardless of the source. Instead, it entailed Ghanaians adopting coequal status with other staff colleges, considering Ghana a legitimate partner in providing military education and training for international officers, and assuming an intellectual cohesion based on particularly Ghanaian experiences.

Ghanaianization was also Africanization. Another former faculty member from the late 1980s recalled, "The idea of the staff college, the whole idea of the Ghana Armed Forces Staff College was started by Osagyefo Kwame Nkrumah. You wouldn't believe it, he started it. . . . The idea was to train Ghanaian and other African countries."[241] By 1989, he recalled, "when you go to our staff college, you have Tanzanians and Ugandans there, and in fact, nobody pays anything. All those countries, they don't pay anything. You see, Ghanaians have that attitude, our sister countries let us help them. . . . So you have a police man, you have a fireman, you've got a Nigerian, You've got somebody from East Africa . . . there was a man from Rhodesia." Throughout the 1980s, Ghana and Nigeria expanded their student exchanges and increased the number of faculty exchanges between the Ghanaian and Nigerian staff. Meanwhile, it accepted officers from Tanzania, Zimbabwe, Uganda, Sierra Leone, and even two officers from Transkei in 1992, just prior to that state's reabsorption into South Africa.[242]

Ghana's official historians and the college's former faculty remembered the staff college as the fruit of broad international military cooperation, an indigenous institution of higher learning in the service of the national military and a pan-African source for professional development and education. Kwame Nkrumah began the staff college idea as a Ghanaian initiative to serve pan-African ends, but throughout its long period of gestation, the staff college could never remove itself from global trends in international military educational exchanges—nor did it ever wish to. Nigerian students joined the staff college's third class in 1979; Sierra Leoneans and Tanzanians joined the fifth course in 1981. Neither was its national character a given: during the mid-1970s, despite an agreement with the Ahwereasehene—sealed with six bottles of schnapps and one carton of beer—that the new staff college would be built sixty kilometers north of Accra near Awerease village, local politics and economics dictated that the existing barracks at Teshie would become the school's permanent—while officially temporary—home.[243] Meanwhile, British, Canadian, and Nigerian officers complemented the staff college faculty. So also did Ghanaian officers who had attended staff colleges in Pakistan, the United States, United Kingdom, Canada, and India.

Ghana's staff college evolved at what Jose Moya calls "the meeting of the extremes: global forces and local conditions, of the world and the village."[244] Even in the late 1980s, as the faculty strove to "Ghanaianize" the syllabus and faculty, the staff college never rejected internationalism at the expense of the local. Here again, in Moya's words, "the nation-state may offer the optimal unit of analysis" to study international policies, but it is "a faulty one for examining the actual process."[245] Instead, what Ghana's officers called "Ghanaianization" appears even more internationally integrated than before, only with fewer non-African labels.

## Conclusion

> *If we Africans are serious about achieving our national emancipation, we must first of all have it clear in our minds that we are capable of surviving on our own; that all it would need is proper utilization of the natural and human resources within our countries. Any other assistance from any other source should be considered of secondary value. The harrowing Amin years should sometimes teach us a few lessons. . . . A good part of the world ignored Uganda then. Aid only came in trickles. In spite of the atrocities during the noted years, Uganda boomed back. If the country was able to survive under anarchy, what would have been a situation under organized and enlightened leadership? Surely it would have been better.*
>
> – Major Benon Tumukunde, Uganda, in *The Flame: The Magazine of the Ghana Armed Forces Staff College*

Ghana's leaders accepted grants for military training from around the world for different reasons. On one hand, when President Kwame Nkrumah and the military officers who succeeded him sought education and training assistance abroad, they were filling a specific need for technical assistance Ghana did not yet have the capacity to provide for itself. Nkrumah and military leaders after him also publicly accepted that global norms for military professionalism and ideal civil-military relationships made the military a "modern" institution in society. But as the case of the three lots of textbooks donated to Ghana Military Academy in 1961 demonstrates, political considerations always hung over internationally available education and training resources. How internationally trained military personnel interacted with the Ghanaian state after independence reveals the contradiction between global ideals of the professional military embodied in Military Internationalism and loyalty to the nation inherent in post-colonial nationalism.

Programs that sponsored military education and training for potential leaders became the cornerstone of the United States' security assistance relationship with Ghana because they merged assumptions about the modernizing power of transnational education with the political challenge of asserting US leadership in places where decolonization limited the United States' freedom to intervene. Sponsoring military education and training offered a flexible, expandable response to Ghana's needs that did not also tax the United States' relationship with its other allies. In 1963, the Office of the Director of Military Assistance's five-year plan for Ghana coincided with the secretary of defense's expansion of the Informational Program worldwide. Both plans affirmed the long-term, cost-effective value of international education and training alone. Even after the 1960s, when scholars and policy makers no longer assumed that political and economic development occurred in sequential stages dependent first upon the psychological modernization of a core of elites, the actual operation of American military assistance to Ghana changed little. When the products of American international military education and training seemed to both fulfill and contradict the ideals, as in the case of Colonel Ignatius Acheampong's rise to power and violent demise, the ideology and rhetoric underwent surprisingly little change. Meanwhile, military rule persisted, and Ghana's economy struggled.

Ghana's participation in international military education and training never implied contentment with dependence on expertise it could not provide. From the start, Nkrumah intended Ghana to lead a self-sufficient, pan-African security coalition, and he expected Ghana's military schools to instruct soldiers from across Africa. His aggressive participation in United Nations peacekeeping in the Congo represented that ambition and caused tension with Britain and the United States, whose neocolonial meddling he blamed for Brigadier Joe Michel's airplane crash in 1961. Ghana Military Academy and Ghana Armed Forces Command and Staff College sustained Nkrumah's dream, but the armed forces' effort to "Ghanaianize" the staff college demonstrated the fluidity of roles between consumer of international military education and provider. By the late 1980s, as the epigraph above suggests, a Ugandan military officer attending Ghana Armed Forces Command and Staff College could use the college's annual journal to express his recipe for African independence from the "Colonised Mind."[246] Ghana continued to accept grant aid to send promising military officers to staff colleges around the world, but through its own staff college it offered professional education to African officers as well.

When military rule came to Ghana, the assumption that military education and training yielded psychological results over the long term insulated the United States' specific education and training program with Ghana from significant ideological scrutiny. The United States' overall education and training program survived fierce congressional inspection in the late 1960s for the same reason. Western observers did not initially consider Ghana's foray into military rule in 1966 a fatal blow to constitutional rule or the death of civil-military ideals, but subsequent coups in 1972, 1979, and 1981 obliged American assistance planners to reconsider their programs' intent and capacity to help Ghanaian society. As we shall see, the United States adapted its rhetoric when describing the specific results military assistance could achieve in Ghana, but even great political change in Ghana never undermined the fundamental assumption that international education yielded individual transformations that would, over time, transform entire societies.

# 3 Writing for Peace

## *International Men and Women of the Ghana Armed Forces*

### Makana's Fanara Diary

In February 1975, as his six-month peacekeeping deployment came to an end and he and his battalion of five hundred Ghanaian soldiers prepared to leave Egypt, Lieutenant Makana Blay published his observations and impressions in *Ghana Armed Forces News*.[1] His full-page article described the Ghanaians' mission, the terrain, sandstorms, Cairo and its charms, and positive and negative interactions with Egyptian Arabs. Beneath a photo of four Ghanaian soldiers and a dozen or so children, his caption read: "The soldiers hardly dine without their newly acquired friends (Egyptian Children) coming around for their share. And this is well within the spirit of African Unity: Be Thy Brothers Keeper."

Three "lessons" from the experience struck him:

"Our men can talk face to face with other nationals without any feeling of inferiority complex. They know that they are and can be as efficient as any other soldier from any part of the world."

"We have seen how other nationals live. Some . . . are less better off than the Ghanaians but they don't complain much. . . . They don't curse the world when there is shortage of sugar and milk."

"As for Pte [Private] ABONGO he now knows that a dollar is a paper note just like the cedi. When he was first paid his dollar allowance popularly known as a 'Kissinger' he uttered in surprise, 'so the dollar way people de talk about ibi [it be] paper? I tink say ibi [it be] gold.'"[2]

Since the earliest days after independence in 1957, members of the Ghana Armed Forces and their families consciously and intentionally joined a global community of military service best described as internationalist. Chapter 2 argued that civilian and military political leaders routinely pegged Ghanaian military education and training to international norms (usually Western, but not exclusively) and restated national military objectives using internationally recognizable reference points and language. Even taking these leaders' statements at their most cynical and self-serving to justify military coups and regimes, they resulted, in practice, in actual international travel for thousands of Ghanaian military men and their families in the 1960s, 1970s, and 1980s.[3]

Instead of searching the Ghana military's failure to fulfill national ideals or its limited success supplying a locus of nationalism around which Ghanaian identities coalesced, this chapter looks for alternate uses of military service for individual Ghanaians' identities. It argues that between the 1970s and 1980s, Ghanaian military personnel and their families found the experiences of international military education and peacekeeping personally rewarding but that travel also connected Ghanaians to global communities while weakening some national bonds. International military service provided Ghanaian families alternate strategies to negotiate economic insecurity in ways that strikingly resemble other diaspora communities, with an essential difference: in this case, the transnational identities of the families of Ghanaian soldiers still depended on functioning state agencies and international diplomatic processes to facilitate their travel, and the state usually expected them to return promptly after their training. These Ghanaians' alternate identities were both family-centered and global, employing international travel as both an economic and political strategy to "get ahead" and to forge global relationships based on an international humanitarian ethos.[4] This brand of internationalism transcended the state but still depended on military scaffolding only the state could provide.

Especially during the period of Ghana's military rule (1966–1992), historians had struggled to reconcile the possibility of an "imagined community" of Ghana with the evidence of massive political, economic, and cultural fractures in the Ghanaian state.[5] Ghanaian historiography continues to grapple with whether Ghana met "national" criteria at all, as ongoing competition between regional and ethnonationalism makes nation-state status fit uncomfortably around Ghanaian history in the three decades after independence.[6]

Not even the military, the proto-national institution that actually administered the state for most of those years, remained free from society-wide cleavages between corporate and class identities, elitism and radicalism, and ethnicism.[7] On the contrary, involving themselves in politics made Ghanaian military leaders no more likely to put national interests above regional, ethnic, or corporate ones.[8] From the "right-leaning" coup in 1966 to the "left-leaning" one in 1972 to the "revolution" of 1979, at one stage or another, all leading social factions in Ghana imagined their class interests conjecturally reflected in the military.[9] Likewise, scholars generally agree that the Ghanaian armed forces, like most others in sub-Saharan Africa, failed to represent the ideals of "professionalism" that were supposedly necessary to bring political development.[10] Cloven as it was by class, ethnic, and political divisions, rarely, if ever, did an organization called "the military" act in a "national" way.

In the same ways they complicated Ghana's already fractured political economy after independence, persistent ethnic and regional allegiances complicate any discussion of class within the Ghanaian military. Other historians have more closely examined the fractured allegiances military members forged across class lines.[11] This chapter, therefore, employs the term "class" to describe the socioeconomic patterns in Ghanaian military families' income and consumption that connected them to similar communities of soldiers in Ghana but that also translated—however inexactly—to international economic standards for soldiers modeled in places like the United States. This chapter argues that the thin cultural barrier between Ghanaians and the members of other military communities they encountered abroad reflected both sides' misperceptions of the other's class status. Ghanaians equated this perceived difference in status as a disparity in "modernity," and they employed the gap as a strategy to make claims for pay and benefits from the Ghanaian government and from the international community.

Compared with other opportunities available to Ghanaian elites, military service offered some enviable opportunities—especially in international travel and training. Chapter 2 explored Ghana's ongoing relationship with the US international military education and training program. Ghanaian military elites also aggressively pursued inclusion in UN peacekeeping operations, first in the Congo from 1960 to 1963, then in the Sinai between 1973 and 1979 and Lebanon continuously after 1979. After 1989, Ghanaian soldiers or police joined multinational peacekeeping missions around the world—twenty-three in just the next decade—providing eighty thousand soldiers in forty years between 1960 and 2000.[12] Extensive military participation abroad reaped corporate benefits and national ones. The United Nations reimbursed the government of Ghana for equipment depreciation and peacekeepers' salaries, for example. But peacekeepers themselves also developed international relationships while earning significantly more money on deployments than they could in Ghana.

International service altered Ghanaian families as well. Especially on professional courses abroad, wives and children often joined in the travel. In most cases of UN service, deployments abroad forced soldiers' wives and children to adapt to difficult separations as Ghanaian women negotiated with the ideal of the "modern military wife." As a result, in the 1970s and 1980s, military education, training, and UN service abroad—and the political and economic benefits flowing from them—became an essential aspect of the individual identities and economic strategies of Ghanaian service members and their wives.[13]

This chapter also argues that internationalism *itself* motivated Ghanaian soldiers and their families. When abroad on courses or in Egypt or Lebanon in the 1970s and 1980s, members of the Ghana Armed Forces considered themselves part of an imagined *global* community of military scholars and peacekeepers that transcended or exceeded their imagined *national* community as an essential marker and motivation for their service.[14] Soldiers' dispatches from the field, their poems, messages from home, memoirs, even Makana's photo of Egyptian children crowding Ghanaians soldiers at mealtime and its caption proclaiming the "spirit of Africa" evoked this community. To complicate matters, as the nation-state dissolved or reformed in their absence, as it did periodically in the 1960s, 1970s, and 1980s, Ghanaian soldiers—like other Ghanaians—continually renegotiated their *national* identities. Polarizing the issue into international, national, and subnational identities does not imply that these individuals lived more than one social reality. But de-emphasizing the nation for a while reveals ways that other identities expressed themselves, sometimes with surprising intensity.

If rather than looking at the *national* as the essential attribute of military service we instead consider it an occasional, and occasionally uninvited, guest in individuals' much larger worldviews, individuals show themselves constantly negotiating the extent to which they permitted the state to contribute to or hinder larger plans. Ghanaian service members and their families sometimes engaged the state as an economic enabler, sometimes as a foe subjecting them to danger, sometimes as a provider of refuge from danger, or sometimes as a mode of transportation into a global community. Both men and women leveraged military service and the new template for modernity it imposed to make claims for themselves against the Ghanaian state, just as Ghana did to make claims against the international community. Women on the home front learned new ways to manage households in their husbands' absence, for example. They did so by combining some army-sponsored resources with new and existing social networks in combinations that resembled women's strategies in other military communities around the world. Ghanaian soldiers and their families moved easily in and out of the global community, even if—as Private Abongo's realization that the "Kissinger" was a paper note just like the cedi demonstrates—the initial international experience was jarring.

Part of the definition of the post-independence African diaspora includes the ability to move freely between family, ethnic, national, and international identities.[15] After independence, individual Africans, especially Ghanaians, made transnationalism and travel key strategies for survival and accumulation. According to historian Emmanuel Akyeampong, for Ghanaians, "going abroad" meant "getting ahead."[16] International military structures provided Ghanaians vast numbers of opportunities for "going abroad," so it makes sense that the transnational or international experiences they made along the way altered their local and national identities. Continuing pre-independence patterns of mobility, the combination of voluntary and reluctant military travel also resembled other Africans' enthusiasm and hesitation to join the diaspora.[17] A closer look at international military service, therefore, offers striking similarities—with some significant differences—between Ghanaian military personnel and their families and other African diaspora communities abroad.

The chapter begins by exploring Ghanaian individuals' enthusiasm for international military travel in the context of the corporate and national benefits it also provided for the Ghana Armed Forces and the Ghanaian state in the 1970s and 1980s. Much of the political uncertainty that threatened Ghanaian political leaders also affected Ghanaian soldiers—which is not surprising because they were often the same. For Ghanaian military families, international travel for education or UN service provided temporary economic or political asylum, or offered benefits that loosened the restraints of class and that they could parlay into economic independence in the future. Military families often used extra pay and benefits to build houses for themselves and furnish them with appliances that were scarce in Ghana. Thus they bridged traditional values governing domestic space with recent innovations that represented their changing criteria for the modern home and modern family. While abroad, Ghanaian families also formed global networks with other military families, with other soldiers, and with the people whose peace they deployed to protect. UN service was the most common method for forming these networks and building economic and political buffers, but UN service also caused lengthy separations between Ghanaian soldiers and their wives. Women, therefore, participated in the internationalist identity differently than men, often by becoming more independent and detaching somewhat from prior regional or group identities. Finally, the chapter explores international military service not as an economic or political strategy, but as a transnational identity. Considering its ongoing dependence on state resources, this identity is best described as international, based partly on national pride but largely on equal participation in a global project for peace.

Between the mid-1960s and early 1990s, the military dominated Ghanaian politics, but this chapter does not dwell on military rule or its impact on Ghanaian society or the armed forces themselves. With few exceptions, historians concur that military rule was disastrous for Ghanaian political culture, Ghana's economy, and even the armed forces' own development.[18] In the late 1960s and 1970s, Ghana's military declined in capability and morale as precipitously as the rest of Ghana's economy. Some sympathetic voices suggest that military rule was a symptom of larger regional and global stress and that the armed forces declined in parallel to Ghana's general economic decline after 1962.[19] None claim that military rulers delivered on their promises to restore national unity, end corruption, or deliver prosperity better than a civilian government did, or could do. Neither does this chapter. Instead, it attempts to evaluate how individual families in the Ghana Armed Forces employed the military's tenuous, but ongoing, connections to international military exchange and employment structures in their personal strategies for political or economic liberation and how they internalized the moral aspects of participating in a global military structure.

## "We know the sea that washes our shores": The Politics of International Service

So far, this work has examined the United States' International Military Education and Training Program from the perspective of US foreign policy, from the perspective of American communities, and as an essential element of the Ghana Armed Forces personnel policies. For Ghanaian families, international training or peacekeeping deployments abroad also offered a method to mitigate political and economic uncertainty at home. Much of the evidence describing the way individuals employed international military travel between the early 1960s and 1990s suggests that international travel altered the way Ghanaians identified with their own state. Like other Ghanaians, military members were also vulnerable to changing political and economic circumstances in Ghana and abroad. Military service combined the two conditions—opportunity for international training through travel, and vulnerability—in ways that were both common and unique among other groups of Ghanaians abroad.

* * *

Ghana Armed Forces leaders had always touted military service for offering opportunities for advanced education, but recruits especially responded to the opportunities for education and training abroad. Even before independence, promising cadets attended the Royal Military Academy at Sandhurst. Six left for England in 1956; two did in 1957. Thirteen more, including G. Dako, did in 1958.[20] By 1970, sixty Ghanaian men had gone to Sandhurst.[21] The chance to attend Sandhurst inspired other young men from Dako's hometown of Achimota to join the army as well, he recalled, which was significant because Akans from Achimota previously reviled military service.[22] Dako's father "protested vehemently about his joining the army." At Sandhurst, Dako met cadets from Iraq, Malaysia, South Africa, Kenya, Uganda, and Sierra Leone.[23] He later remembered his service as "a world of extensive travel throughout Ghana, Europe, and America."[24] An army recruiting advertisement in the *Ghanaian Times* promising "overseas training in the United Kingdom" convinced Henry Kwami Anyidoho to compete for a job as an army radio operator in 1960.[25] "In those days," he recalled, "to travel to the United Kingdom alone was a great achievement in society." International travel, in other words, permitted Dako and Anyidoho to transcend former regional and ethnic prejudice against military service. Shortly after the 1966 coup, in his *Ghana Armed Forces News* article celebrating the putsch, public affairs officer Lieutenant A. A. Enninful boasted, "A number of young officers are currently on study leave in Ghanaian universities. Elsewhere in European universities there are lots more studying engineering or doing post-graduate courses."[26] For Ghanaians, Emmanuel Akyeampong argues, "to travel is to get ahead," and for some, military service offered a promising opportunity for just that.[27]

Since independence, international education programs were necessary for promotion to higher ranks in the Ghana Armed Forces, they bestowed prestige, and military and political leaders manipulated them as rewards. Like any other political prize, overseas courses could also be withheld from political enemies. American ambassadors knew it; so did the Defense and State Department officials who ran the United States' international military education and training program.[28] Just before his overthrow in January 1972, for example, Prime Minister Kofi Busia attempted to defuse a perceived threat to his office by ethnic Ewes by scattering Ewe officers, sending Air Force Commander Brigadier Ashley-Lassen to the Imperial Defense College in India for a year.[29] That plan failed. Colonel Emmanuel Erskine, one of the nine members of the ruling National Redemption Council junta who ousted Busia, accepted a one-year posting to the Imperial Defense College in England just three weeks after the coup as a reward.[30] Recalling his staff college experience three decades later, one former Ghanaian officer was convinced that "his politics" had both led to his selection for the US Command and General Staff College in 1976 and later capped his career at brigadier in the 1980s.[31]

In the wake of the 1979 coup, the general fractures in the Ghana Armed Forces became tangible to Ghanaians struggling to secure or retain overseas courses they thought they had earned. Henry Kwami Anyidoho's experience is a case in point. In spring 1979, Anyidoho "was all joy" when he learned of his selection for the staff college at Fort Leavenworth.[32] During the June 4, 1979, uprising, however, junior officers and noncommissioned officers loyal to Flight Lieutenant Jerry J. Rawlings detained then-major Anyidoho along with most of Ghana's other senior officers for questioning and trials. His regiment's junior officers locked Anyidoho and his colleagues in the headquarters coffee room for several days before transferring him to Flagstaff House, Ghana's capitol building, to join senior officers from across Ghana in detention.[33] At military bases elsewhere, junior officers and men incarcerated the senior officers in the Officer's Mess, prohibited them from receiving visitors or food from their families, and shaved their heads.[34] By the end of June, Rawlings's Armed Forces Revolutionary Council (AFRC) released most of the senior officers, but executed eight Ghanaian generals, including former Head of State Acheampong, the chief of defense staff and chiefs of staff from the army, air force, and navy.[35]

The junta eventually released Anyidoho, but "a young signal officer" who Anyidoho had angered earlier in his career derailed Anyidoho's posting to Fort Leavenworth. "My course was cancelled and restored, then cancelled again and restored, and cancelled," he recalled.[36] "The mental torture was so much that I was losing weight rapidly."[37] Anyidoho decided to forget about the course, forget about the army, and concentrate on building up his family farm. "My children were growing so I farmed to supplement my pay in order to feed them properly," he wrote.[38] In fact, nearly every Ghanaian military family supplemented its income in some way during the 1970s and 1980s by farming, cooking, or performing additional domestic work for pay.[39] During the heyday of "Operation Feed Yourself" in the 1970s, military units themselves also ran cattle ranches, poultry farms, cassava farms, rice, and yam farms on military bases across Ghana.[40] Anyidoho had used the income he and his wife Mercy had earned living in the United States in 1975 and 1976 to begin a palm oil plantation near Accra, and he retreated there after his humiliation in June 1979. "It was when I returned from the Pokuase farm one evening," he recalled, "that a young officer from Signals, Captain Egeme-Ari had come to look for me; I was to leave for the course the following day! I was not enthusiastic anymore but decided to think it over."[41]

Anyidoho had trained in the United States twice before. As an unmarried corporal, Henry became the first Ghanaian to attend a technical school in the United States in April 1962, when he completed the Field Radio Repair course at Fort Gordon, Georgia. Despite firsthand experience with Southern racism—he was thrown out of an off-post pool room because of his color—he still considered the trip "one of my life dreams came true; I landed in the white man's country!"[42] "A good deal of integration had taken place" in the American military, he recalled, and he felt "very lucky to have established a very good relationship with most of my white classmates."[43]

Then-corporal Anyidoho did not dwell on the experience of being a black African in segregation-era Georgia. In his memoir, he did not mention it again, even when he came back to the American South in 1975 and 1979. In May 1975, Major Anyidoho returned to Fort Gordon for the yearlong Advanced Communications Course.[44] Mercy, who had a certification from the Nursing Council of England and Wales, joined him in Georgia and found a nursing job at a hospital in Augusta, even though she had given birth to their daughter Mawuse just three months prior. Mercy's mother moved into his on-post house in Accra to mind Mawuse. In Georgia, meanwhile, Mercy cried "night after night" over the little daughter she left in Ghana.[45] Henry and Mercy made friends with other Ghanaians in Augusta, including a doctor from the hospital and his wife, two other Ghanaian officers at Fort Gordon and their wives, and the wife of a Ghanaian officer who had stayed in the United States "to do a degree course in Nursing" after her husband finished his course and returned to Ghana.

In Georgia in 1976, Henry and Mercy invested their money carefully. "Georgia being a cotton producing state," he wrote, "we bought a lot of clothes for our children and our friends. . . . Most impressive of all, we bought two new Toyota cars."[46] They planned to keep one, he recalled, but the other "was to be sold in Ghana for us to start a private housing project. A turning point indeed!" When Henry's course ended and he returned to Ghana, Mercy stayed behind for a few months "since she was doing a short course and also wanted to earn a few more dollars." He remembered, "We were going to miss each other for a while, but for a good purpose."[47]

Henry and Mercy's experience in Georgia closely resembled other economic migration strategies for nonmilitary Ghanaians of the same period. African émigrés often found work in hospitals and clinics in nonmetropolitan communities in the United States during this time.[48] Ignatius Achel's wife worked at a nursing home in Kansas while he attended Fort Leavenworth's Command and General Staff College, for example.[49] In a kind of mini-chain, Henry first established a home in Georgia, then sent for Mercy, who quickly found a job and remained in the United States after he returned to Ghana. Mercy's arrival inspired the other two Ghanaian officers' wives to come to Georgia and join their spouses, Henry recalled.[50] She facilitated their integration in the host society while maintaining a connection with home, just as transnational African immigrant networks do elsewhere.[51] Despite great sadness leaving her children, Mercy understood her contribution to the migration, saw herself as Henry's partner, and even remained in the United States to maximize the family's economic gain.[52] The Anyidohos arranged for extended family members in Ghana to perform child-rearing tasks in their absence, but they also made use of the resources of the Ghanaian state—in this case, Anyidoho's on-post house—to facilitate the process.[53] Finally, the Anyidohos used their savings from the migration to build a home and business in Ghana, fostering a transnational identity that straddled multiple societies.[54] For Ghanaians in the diaspora, building a home expressed material well-being, dedication to ancestors and extended families, and affirmed ties with the homeland.[55] In short, during their year in Georgia in 1975 and 1976, the Anyidohos capitalized on a state-to-state program for military cooperation to execute their own family's transnational economic strategy while maintaining a simultaneous "presence" in home communities as well.[56]

Their experience with staff college in 1979 revealed a different purpose for international education. The June fourth uprising fractured the Anyidoho family. Mercy and the two youngest children moved to her brother's family home. Their two older sons lived with Henry on-post at Burma Camp. "I brought the bigger boys Akpe and Kafui to sleep in the Camp for two reasons," Henry recalled. "Firstly, I wanted them to grow up as strong and bold men who will not panic at the slightest sign of danger and secondly just in case I got picked up in the night, they could be witnesses to it."[57] After receiving word that Henry's staff college posting had returned, Mercy convinced Henry not to pass up the opportunity. "Her reason was simple," he remembered. "The army was my chosen career so I should pursue it."[58] When a driver arrived to take him to the airport the next morning, he had still not told his children he was leaving. "What kind of life was this!"[59]

Throughout the trip, he remembered, his thoughts returned to his wife and children: "Were they safe?"[60] Henry discovered mid-journey that he was headed to the US Marine Corps Command and Staff College in Quantico, Virginia, and not Fort Leavenworth. He became only the second Ghanaian officer to attend there.[61] "I did not want to leave anyone at home because of the uncertainties," he wrote, so he applied through the Ghanaian embassy for Mercy and the children to join him in Virginia, borrowing money from friends and relatives to pay for their travel.[62] When they finally arrived, he remembered, "all the wives of the other officers in my den, and we were twelve in all, came to my bungalow and put the house in order."[63]

The Anyidohos' year at the Marine Corps staff college in the United States kept them safe. They had leveraged the Ghana Armed Forces' continuing relationship with international training and education programs around the world for their own protection. In a time of danger after the coup, the Anyidohos manipulated the Ghana government's longstanding international commitment to obtain protection from violence the domestic Ghanaian state could not contain. They found in the American community around Quantico Marine Corps Base a welcoming multinational community that was familiar with international military officers and that the Anyidohos easily recognized from their prior experience in the United States for military education.

The Ghanaian state initiated and administered international military education, but individuals abroad continually negotiated with the state's uncertainty. In January 1980, Major Michael Abana readily accepted a short-notice assignment to the Pakistani Command and Staff College in Quettah despite the army's promise that they would send him to the Indian staff college later in the year. Given the political upheaval of the prior year in the Ghanaian armed forces, Abana said, he feared he might never get the opportunity to attend staff college if he passed up that chance.[64] By then, he said, "I was fed up with military rule in Ghana for a while yet" but had no illusions about Pakistan's ongoing military governments, either.[65] "There was a day that we were discussing these sorts of things in the classroom, when I told [my Pakistani colleagues] point blank that it was not their responsibility to rule, they are to soldier!" he recalled.[66] "I was lucky I was not sacked! Why could I have been sacked? [Because] they have told us, 'we know the sea that washes our shores, but we do not talk about it.'"[67] They need not mention the Indian Ocean by name; they need not mention the illegitimacy of military rule. Both just were. Both the Pakistani army and Abana understood that military rule was problematic, but with a combination of fatalism and resolve, the staff college continued, and continued to accept Ghanaian officers. Meanwhile, Abana and his wife Peggy both learned a different perspective about Pakistan. "Even up till now," he mused thirty years later, "whenever something happens in Pakistan, if she hears, she says 'Hey Mike, this is happening in Pakistan.' It was a good education for her, too." Michael Abana, Peggy, and his Pakistani colleagues all considered their friendships legitimate, even if they knew their respective governments were not.

Sometimes the Ghanaian state did worse than just frustrate the military scholar abroad. Like Henry and Mercy Anyidoho—but on the other side of the world—Lieutenant Steve Oduro-Kwarteng attended a signals officer course overseas in India, in 1975. Like the Anyidohos, he saved his money. "At the end of the course, I was able to return to Ghana with a dashing two-door Datsun car ordered directly from Japan," he wrote.[68] Similar to Anyidoho, he left Ghana again in autumn 1979 for a three-year program in electrical and telecommunication engineering at two military universities in India. "In the second year of the degree course," he recalled, "I received a confidential letter from the defense advisor at the Ghana high commission in India telling me of my retirement from the Ghana Army." In his absence, the Ghana Armed Forces had decreed that all officers and senior enlisted personnel who were "closely associated" with Rawlings's Armed Force Revolutionary Council "should be retired from the Armed Forces."[69]

Oduro-Kwarteng lived through the January 1972 coup as a junior officer, and he understood the perils of a politicized military, but he could not avoid the chaos of Flight Lieutenant Rawlings's June 4, 1979, revolution. As the coup began, Rawlings and his AFRC detained all officers with the rank of major and above and directed the senior captain in each armed forces unit "to take control of affairs in the unit until things were properly sorted out" and instructed "that no senior officer was to be touched."[70] This order triggered Major Henry Anyidoho's captivity in the regimental coffee room at the Ghana Armed Forces headquarters at Burma Camp. At the first major meeting between Rawlings's council and representatives from across the services, Rawlings selected then-Captain Oduro-Kwarteng to take notes and produce a transcript for the meeting. Oduro-Kwarteng served as Rawlings's recorder routinely after that, often appearing in photographs beside Rawlings, although Oduro-Kwarteng claimed he had no ideological stake in the revolution.[71] Rawlings retreated from power temporarily between September 1979 and December 1981, and Ghana's new civilian regime attempted to restore the military hierarchy by offering AFRC members the choice "to go on courses overseas, to retire and be resettled, [or] to take extended leave before re-joining their units."[72] Oduro-Kwarteng, already on a course in India at that time, shared Rawlings's (temporary) fate: the Limann government forcibly retired both from the Ghana Armed Forces.[73]

A formal international agreement had sent Oduro-Kwarteng to India in September 1979, but he quickly exercised his own initiative when the scaffolding of national sponsorship dissolved beneath him. In May 1981, the Ghanaian government formally "retired" him, but the Indian military university he attended did not acknowledge his changed status, so Oduro-Kwarteng kept the news to himself and "concentrate[d] on the course."[74] "Because I was no more recognized as an Army Officer," he wrote, "I told myself that if the Ghana government did not want me in the Ghana Army anymore, the Engineering market in the civil world might want my services. From then on, I forgot about the Ghana Army and rather worked harder on my books and projects."[75] He finished the course in 1982 and earned a bachelor's degree in engineering, but his family faced a difficult choice.

Steve Oduro-Kwarteng, his wife Genevieve, and their three children (one born in India) were out of work and out of money. "Some universities were prepared to accept me for the Masters Degree," he wrote, "but it was financially very difficult for me to sponsor myself in a university in India."[76] They considered staying in London, "but [we] did not have the resident permit."[77] They knew that "back at home in Ghana, people had been looking for AFRC members and their associates to bring them to face justice." In fact, Rawlings had overthrown President Limann in December 1981, and his own Provisional National Democratic Congress (PNDC) was behind the purge of former AFRC members. Nevertheless, Oduro-Kwarteng recalled, "I had nothing to hide and nothing to be afraid of, so I did come home."[78] In 1983, the Ghana Army reinstated Oduro-Kwarteng and restored his captain's rank. When the state retired Oduro-Kwarteng from the service, he "forgot about the army" and concentrated on his studies as an engineering student. After weighing their options and returning to Ghana, Steve and Genevieve warily reestablished their connection to the army and the Ghanaian state.

In both the Oduro-Kwarteng and Anyidoho families' cases, the Ghanaian state facilitated the international travel that they wielded in their own defense against the state itself. Neither dwelled much on this contradiction in their memoirs, but both recognized that their service in the armed forces was contingent on political factors in Ghana that they could not control. Kwame Anthony Appiah argues that by the mid-1970s, most ordinary Ghanaians avoided interaction with army officials, who so dominated Ghana's civic and economic life that they were more likely to be "feared and resented" than welcomed.[79] Both of these families maximized the protection the state offered while simultaneously insulating themselves from the danger it threatened. This tense relationship with the state, which shifted between national military service and transnational labor migration, resembled the experience of other Ghanaians in the diaspora, who often considered their migration "a voluntary and personal act undertaken without government role."[80] Ghanaian military families usually returned to Ghana after shorter absences and relied on formal state mechanisms to execute their travel plans, but by joining Ghanaian communities abroad, investing their pay carefully, and building homes for themselves on their return, they insulated themselves from the state's unpredictability in ways similar to others in the diaspora.[81]

International military service with the United Nations offered Ghanaians other opportunities for similar self-preservation. Major General Emmanuel Erskine employed a related tactic in June 1979, when he carefully negotiated for three hundred of Ghana's contingent of five hundred peacekeeping soldiers in the United Nations Emergency Force II (UNEF II) in the Sinai to move to the new United Nations Interim Force in Lebanon (UNIFIL) rather than return to Ghana as the UNEF II mandate ended.[82] Erskine considered service in the bloody Lebanese war safer than returning to Ghana. "I felt that to send the troops home might act as a catalyst for exacerbating the already difficult and uncertain situation in Ghana," he later wrote.[83] Like him, many of the soldiers may have been in personal danger as well. In February 1972, just weeks after Ghana's second coup, Colonel Acheampong rewarded fellow National Redemption Committee member then-colonel Erskine for his three weeks' service as army chief of staff with a one-year posting to the Imperial Defense College in England.[84] Erskine resumed his position as army chief of staff after he returned in January 1973. In November 1973, Erskine led the Ghanaian delegation that negotiated with UN Secretary-General Waldheim for Ghana's inclusion in the UNEF II peacekeeping mission and led to his own selection as chief of staff for the UN force.[85] Erskine served exclusively with the United Nations in Egypt and Lebanon after that, commanding the UN Interim Force in Lebanon until 1981 and commanding the United Nations Truce Supervision Organization until 1986, when he finally returned to Ghana.[86]

International military service kept Erskine safely away from Ghana during the turbulent years surrounding the demise of the Acheampong regime and the rise of Rawlings without ever damaging his reputation. The stakes were high. The three army commanders who succeeded him were killed in the coup or executed in June 1979, and Erskine understandably avoided Ghana during that time.[87] He specifically canceled a planned trip to speak to the Ghana Armed Forces Command and Staff College in June 1979 and spoke at the Nigerian Staff College instead.[88] Erskine convinced the United Nations to pay him directly at a pay scale comparable to other UN officials and far higher than the Ghanaian standard so he could afford for his wife and six children to remain in England while he served in the Middle East.[89] Throughout the 1970s and even during the Rawlings years, *Ghana Armed Forces News*, the Ghanaian press, and the Ghanaian embassy celebrated Erskine's achievements and visibility as a highly decorated, internationally recognized Ghanaian military officer leading a global mission for peace.[90]

For these three military families—the Erskines, Anyidohos, and Oduro-Kwartengs—international military education and UN service offered protection from political uncertainty in Ghana. The three families' responses to that uncertainty also revealed an allegiance to an international ideal of military corporateness that transcended national identities. For them, the international benefits of military service persisted when national structures failed.

* * *

There was nothing surprising or unusual about the wives of the other eleven officers in the Anyidohos' class at the Marine Corps Command and Staff College preparing their home for Mercy's arrival. In practically any other setting, one might expect any other group of neighbors or classmates to do the same. In fact, the gesture's very ordinariness—and the Anyidohos understanding of it—offers hints to the universality of the experience that requires attention. Henry Anyidoho's "den," his community of twelve classmates that worked as a unit through the staff college year, included a Canadian officer, a US Army officer, the Anyidohos' sponsor, and eight other American Marine Corps officers.[91] Two other African officers, Commander Kassem Amir Abdel Hamid of Egypt and Lieutenant Colonel John Shagayah of Nigeria, were in other "dens" at the college. After the course, Henry and Colonel Shagayah remained close friends and visited each other regularly. In the 1980s, during UNIFIL deployments, Henry visited former classmates Major Adnan Bechara of Lebanon and Colonel Dedvani Jehuda of Israel (in different rooms, presumably, but not necessarily).[92] Henry's English professor, Dr. Argus Tresidder, complimented Henry's work, challenged him to improve as a writer, and later inspired him to write *Guns Over Kigali*, his memoir of serving in Rwanda.[93]

International military education connected the Anyidohos to an alternate global community where even spouses' relationships mirrored the classroom's internationalism. Participants publicly sported national military insignia, but only to identify them as a transnational partner, not supplicant. At least Henry Anyidoho thought so. He boasted that his final essay on African unity "won a great deal of admiration" and became a research resource for the Marine Corps Command and Staff College curriculum for many years.[94] Such an identity, constructed on a world scale rather than a national one, did not entirely contradict the nation-state, but it illuminates alternate relationships between families in international military communities.

## Internationalism Accessed through UN Service

Given UN service's clear benefits to the Ghana government's national interests, the military's corporate interests, or even individuals' economic interests, it is difficult to parse the extent to which Ghanaians' expressions of international solidarity with soldiers and peacekeepers represents individuals' internationalism and "modernity." After Ghana joined the UN Emergency Force in the Sinai (UNEF II) in 1973, the United Nations paid Ghana well, and Ghana's military governments passed much of that reimbursement on to soldiers' families. Still, the experience of peacekeeping service required individual families to derive meaningful explanations for the difficulties of the mission, the potential danger of peacekeeping service, and the combination of personal benefits (in the form of increased pay) and global benefits (in the sense of international peace) that peacekeeping accrued. This section describes how UN peacekeeping served national, corporate, and individual needs in order to reveal Ghanaians' expressions of international solidarity in ways the simple terms of nationalism, power, class, or wealth cannot explain. Instead, UN service, like military training and education abroad, actualized in service members and their families an international identity, membership in an unbounded group that employed national resources for its creation and support, but within which individual participants readily exchanged national identities for supranational ones sometimes, and subnational ones other times. The next section will explore more deeply how Ghanaian soldiers made the international act of peacekeeping personal.

* * *

Lieutenant General Emmanuel Erskine, the first Ghanaian soldier to command a United Nations peacekeeping force, affirmed in 1973 and again in his 1989 memoir an international identity accessible through global military service and urged further Ghanaian participation in UN peacekeeping missions to gain access to it. Corporate stresses within Ghana's military also motivated him. Erskine attended commissioning training and two advanced staff colleges in the United Kingdom in the 1960s and 1970s, as well as signals officer training in the United States in 1962.[95] In 1973, as army commander, he claimed that Ghanaian military leaders "yearned" for another UN-sponsored peacekeeping mission to provide experience and opportunities for Ghanaian soldiers, just as Ghana's mission in the Congo had done a decade earlier.[96] The Congo crisis had brought unusually rapid mobility for army officers between 1960 and 1963, but the end of the mission had also caused training, experience, and promotions to stagnate.[97] Simon Baynham, the eminent historian of Ghana's military, argues that service in the Congo and Ghana's period of rapid expansion and promotion in the early 1960s created a massive clot of men with near-equal age, rank, and expectations for advancement despite a dearth of promotion opportunities.[98] "Institutionalized instability" resulted, Baynham argues, partially blaming the 1966 and 1972 coups on the ongoing competition for training, education, and opportunities for promotion that resulted.[99]

Erskine undoubtedly perceived in 1973 that Ghanaian participation in UN peacekeeping duties offered more than just a relief valve for the deadlock in the Ghana Armed Forces' personnel system. Grafting Ghana's military fortunes onto internationally acceptable military missions funded by the international community itself cannily gained the military access to external sources of cash, justified continued claims for grants in international military training and education from abroad, and provided regular opportunities for officers to lead deployed provisional units that did not require restructuring Ghana's own military. As Chapter 2 described, the United Nations required "staff trained" officers to fill high-profile leadership posts in the Sinai and Lebanon, which Ghana was keen to fill.[100] Between 1974, when the first Ghanaians arrived in the Sinai, and the late 1980s, Ghanaian battalions of five hundred to seven hundred personnel rotated into the Sinai (until 1979) and Lebanon (after 1979) every six months. Each battalion included a commanding officer and appropriate staff titles that did not need to match duty titles in Ghana. Continuously after February 1974, Ghanaian officers and noncommissioned officers also filled essential posts in UNEF and UNIFIL headquarters.[101]

Recent scholarship has exposed UN peacekeeping as promising more benefits to contributing states than it delivered, as Ghana's experience in the 1970s partly confirms. Ghana participated in UN peacekeeping projects in the Sinai until 1979 and continuously in Lebanon after 1979, but democratization—an oft-touted symptom of contributing peacekeepers—did not result in the end of military rule in Ghana until 1992. Instead, Ghana's military leaders embraced peacekeeping as a new mission, but did not change state institutions to become more democratic.[102] Likewise, Ghana's well-reimbursed missions in the Sinai and Lebanon may have laid the foundation for other "unintended consequences," including the potential for corruption in equipment acquisitions and in selecting personnel, as Ghana joined dozens more UN and regional peacekeeping missions in the 1990s.[103] On the other hand, some evidence suggests that decades of stable reimbursements have made Ghanaian soldiers reluctant to overthrow Ghana's elected government again and possibly threaten lucrative peacekeeping opportunities.[104]

When combined with significant financial incentives paid directly to the military members, service with the United Nations offered tangible benefits to both the Ghanaian state and the Ghana Armed Forces as an institution within the state. Early in 1974, Ghana's United Nations delegation led negotiations demanding that all UNEF troop–contributing nations—at that time these included Ghana, Indonesia, Nepal, Panama, Peru, Senegal, Poland, and Canada—should receive the same pay, ending the prior practice of reimbursing "poorer" countries less than "wealthy" ones.[105] The United Nation's reimbursement rate climbed from roughly five hundred US dollars per person per month in 1974 to over nine hundred US dollars per person per month in 1980, of which the Ghanaian government returned approximately half to the individual soldiers.[106] This extra pay reached a large number of Ghanaian soldiers and their families. The Ghana Army declined from about 11,000 to 8,000 soldiers between the early 1970s and early 1990s, but 31,000 Ghanaians served tours in either the Sinai or Lebanon between 1973 and 1990.[107] In September 1975, after Nepal and Panama discontinued their service, Ghana agreed to send an additional 200 troops per six-month rotation, increasing their contingent from 500 to 700.[108] For similar reasons, in 1981 Ghana agreed to boost its UNIFIL contingent from 476 to 650.[109] By 1982, Ghana also claimed between US$5.5 million and US$7.5 million per year in equipment reimbursement for its 650-member UNIFIL contingent, nearly half of the Ghana government's own military budget in 1984.[110] Beginning in 1974, the United Nations reimbursed Ghana Airways for delivering Ghanaian personnel and equipment to the Sinai and Lebanon at rates higher than the UN's normal competitive contracting procedures prescribed.[111] Erskine touted UN service as an "immense help to the socio-economic development of the country"; at least, it paid for itself.[112]

Just as the Anyidohos and Oduro-Kwartengs had when on professional education courses, Ghanaian soldiers and their families invested their extra pay carefully in ways that insulated them from the Ghanaian state itself. Like other for Ghanaians in the post-independence diaspora, for soldiers, building a house became a priority.[113] After Steve and Genevieve Oduro-Kwarteng returned to the Ghana Army in 1983, they took a large mortgage and began construction. "I started getting worried that at that stage I did not have even a building plot, let alone a building anywhere for use after retirement or particularly in case of emergency," Steve Oduro-Kwarteng wrote.[114] After peacekeeping deployments in 1986, 1988, 1989, and 1992, they had repaid the mortgage.[115] "Imagine a second lieutenant," suggested retired colonel Michael Abana in a 2012 interview:

> He goes to spend six months. He doesn't buy anything. He brings all the money, maybe $2000 or $3000. What is he going to do with it? He

**TABLE 3: GHANAIAN ARMED FORCES PARTICIPATION IN UNITED NATIONS PEACEKEEPING MISSIONS, 1960–2000**

| UN mission | Location | Duration | Contributing countries | Mission strength | Total Ghana contribution | Mission fatalities | Ghanaian fatalities |
|---|---|---|---|---|---|---|---|
| ONUC | Congo Republic | July 1960–June 1964 | 31 | 19,928 | 8,800 | 250 | 49 |
| UNYOM | Yemen | July 1963–September 1964 | 11 | 189 | 3 | 0 | 0 |
| UNGOMAP | Afghanistan / Pakistan | May 1988–March 1990 | 10 | 50 | 18 | 0 | 0 |
| UNEF II | Suez Canal/Sinai | October 1973–July 1979 | 13 | 6,973 | 8,862 | 55 | 6 |
| UNIFIL* | Lebanon | March 1978–ongoing | 9 | 4,480 | 45,496 | 227 | 22 |
| UNIIMOG | Iran/Iraq | August 1988–February 1991 | 33 | 1,120 | 46 | 11 | 0 |

* Participation was ongoing at the time of publication; data provided here covers the period from 1978–2000.

Source: United Nations Information Center, *Ghana: Four Decades of UN and Regional Peacekeeping*, 26–27. UNIFIL data 1978–2000; "Second United Nations Emergency Force (UNEF II): Facts and Figures," United Nations Department of Peacekeeping Operations, accessed October 6, 2015, *www.un.org/en/peacekeeping/missions/past/unef2facts.html*. UNIFIL contingent strength is the average strength, reported in United Nations Information Center, *Ghana: Four Decades*, 27.

> knows the difficulties in our country. So he will buy a plot of land. And the next time he goes, and he comes back, he will start building a house. So that by the time he becomes a captain, if he doesn't make it [the promotion], he is finished building his house.[116]

After that, Abana supposed, "if you realize that the Armed Forces won't take you any more, you can simply go and resign . . . because of your experience as a military man, you may even get a job."[117] One former chief of defense staff suggested that disappointment with military involvement in politics convinced many Ghanaian officers to quit the armed forces altogether and use their UN allowances to begin businesses.[118]

Soldiers and their families employed their UN allowances to acquire household goods unavailable in Ghana. In 1975, the army chief of staff even commanded them to "spend the extra allowance they would get wisely and return home with enough money to be put in their respective ventures."[119] Lieutenant Oduro-Kwarteng's comrades returned from the Sinai in 1975 with "radio tape recorders, radiograms, black-and-white televisions and tabletop fridges among others, purchased mainly from the Egyptian market."[120] In 1977, the Ghanaian contingent returned from Egypt with a large number of sewing machines.[121] From Lebanon, Ghanaian shipping and import companies facilitated soldiers' shipments of home appliances, corn mills, generators, roofing sheets, and ceramic floor tile–making machines.[122] "For every participation in UN peacekeeping operations," Oduro-Kwarteng wrote, "while some of us were showing interest in acquisition of various types and models of home appliance, a few individuals were interested in saving their allowances for investment in home projects or purchasing commercial items for business."[123] In 2012, retired colonel Abana considered that, in Ghana's dismal domestic economy, the ability to purchase refrigerators and other "essential things" made UN service so lucrative that he thought no Ghanaian soldier would ever again attempt to overthrow the government and risk Ghana's exclusion from future UN service, although Flight Lieutenant Rawlings was not deterred in 1979.[124]

Just as access to international travel unhinged traditional regional and ethnic prejudices against military service, access to external sources of cash also helped destabilize stereotypes of the armed forces as a lower-class occupation. As they did elsewhere in Africa, competing regional, family, ideological, religious, and ethnic identities often subsumed class identities in postcolonial Ghana.[125] Within the Ghanaian military and between the military and the rest of Ghanaian society, those identities competed vigorously, as the ongoing struggle over the Abongo stereotype reveals.[126] In a 1969 letter, a reader who identified himself as "Abongo Boy" complained to *Ghana Armed Forces Magazine* that both on-post and off-post elementary schools discriminated against the children of soldiers.[127] He appropriated the title "Abongo" to file a grievance against class-based discrimination. In 1975, another letter complained about Ghanaian soldiers—"real [A]bongo boys"—cheering against the Ghana Armed Forces' own soccer team during matches.[128] In this case, the letter writer employed the full vitriol of the term to demean soldiers' lack of solidarity—as if the "new" Ghanaian soldier should rise above the stereotype of soldiers as ignorant yokels with no prospects or loyalty. Also in 1975, however, Lieutenant Makana's "Fanara Diary" described "private Abongo's" lesson in world economics and national pride when he realized the Ghanaian cedi and the US dollar were both "paper."[129] Although the stereotype persisted somewhat, by the mid-1990s, Stephen Kojo Addae wrote, Ghana's army barracks "portrayed a lower middle class citizenry, thanks to the many international UN peacekeeping operations that take Ghanaian service personnel to missions all over the world."[130] For Ghanaian soldiers the *international* aspect of military service helped weaken both hard economic class boundaries and soft cultural ones defining military service.

Service with the United Nations provided Ghanaian families with cash and access to household goods, but time apart required wives on the home front to adapt to new social and economic conditions that mimicked older struggles over authority and modernity in Ghanaian family life.[131] Across Africa after World War II, European colonial powers imagined that creating reliable "industrial citizens" required changing prior ideas about work, mobility, property, and family.[132] "Industrial man had a wife," Frederick Cooper writes, and although colonial officials used "masculine pronouns" to describe the transformation to modernity they intended, they directed their efforts at women.[133] "Unless they joined their men–and lived off the men's wages," Cooper argues, women "were the epitome of tribal life and retarding influence."[134] British colonial armies exemplified the process of "detribalizing" male soldiers and their wives, introducing rigid regulations restricting soldiers' marriage, living arrangements, and hygiene in order minimize desertion, maximize discipline, and reproduce future generations of soldiers.[135] Those regulations largely softened in the Ghana Armed Forces, but *Ghana Armed Forces News* still acknowledged that becoming an "army wife" was a process of psychological transformation.[136]

Being an "army wife" required a new identity: "You may say she is a bigamist—sharing [her] husband with a demanding other entity called 'duty,'" Captain K. Adu-Bediaku wrote in *Ghana Armed Forces Magazine* in 1968.[137] "When 'duty' calls, she becomes the No. 2 wife—and until she accepts this fact, her life can be miserable." *Ghana Armed Forces News* described the transformation of Ghanaian women into "army wives" in terms that echo the contradiction between military-imposed modernity and Ghanaian family traditions from the colonial era. American modernization theorists of the 1960s considered extended families—and the financial obligations they entailed—"traditional" and "obstacles" to modernization.[138] Calling on military wives to embrace "No. 2" wife status as a precondition for marital happiness seemed to contradict that aspect of the psychological transformation to modernity. Presumably, being "married to the military" is a common expression for "army wives" around the world, but to call her "No. 2 wife" stings in a special way for Ghanaian women.[139] The emotional transformation required—and its reward of escaping misery—was both distinctly African and universal to the "modern" military family.

The army wife that Adu-Bediaku described defied easy classification. "An Army wife is international," he wrote.[140] She was equally comfortable in the city or the country, and she transcended ethnicity. "She may be an Akim farm girl, a Ga fishmonger, a Kwahu trader or an Army Nursing Sister. When discussing their Army problems, [wives] speak the same language."[141] This description sounded more *national* than *international*, as if military spouses transcended regional or ethnic divisions, and Adu-Bediaku proclaimed such a Ghanaian identity for military spouses. But the identity he described is both smaller and larger than that. He called it "international," but it was also mobile:

> She can be a great actress, watching her children's heartbreak at posting time, she gives an Academy Award performance: "Takoradi is going to be such fun," "I hear they have a beautiful Amenity Center with all sorts of games"; "Farm products and fish are comparatively cheaper. . . " but her heart is breaking with theirs, and she wonders if this Army wife is worth the sacrifice. One day later, *en route* to the new station, and filled with a spirit of adventures, she knows it is.[142]

A similar article from 1975 praised "Mrs. Soldier" as: "A packer, a mover, a chauffeur and a diplomat. Today you are often a member of your national workforce as well. Above all, you're an individual."[143] Adu-Bediaku connected the Ghanaian family to an imagined universal military community that moved freely within and without national borders. As military families often relocated within Ghana—and occasionally around the world—the subnational *family* identity and the supranational *military* identity sometimes eclipsed regional or national ones. Adu-Bediaku intended his article to celebrate army wives' selflessness warmly, not to be a treatise on the militarization of Ghanaian family life, but he also claimed for Ghanaian wives equal participation in the supposedly universal model of army wives around the world, using local idioms to drive home his point.

Women on Ghanaian army bases had always performed unofficial and official domestic duties in camp administration, and they had established support structures even in their spouses' absence. Both women and men routinely supplemented family incomes by farming, cooking, or performing other domestic jobs.[144] Inherited from the Royal West African Frontier Force, the *Sakari Magazia* system promoted influential noncommissioned officers' wives in each military unit with the title *Magazia*, or "woman leader" in Hausa.[145] Typically, the *magazias* supervised sanitation around barrack buildings and disciplined wives and families of enlisted personnel around the barracks. They wore sergeant's rank, accompanied inspecting officers around the camps, and were "so respected and so powerful that any soldier who cross[ed] her could get demoted from his rank to something lower, or fined heavily."[146] During the 1970s, *magazias* from each company and a senior *magazia* from each battalion became advocates on behalf of wives to the larger command and administered the Army Wives Association and Armed Forces Wives Association.[147]

The Ghana Army imagined that women took the opportunity to improve themselves in their husbands' absence on peacekeeping deployments. Instead of normal sanitation duties, "besieging" the paymaster's office, or "the traditional busy cooking around half-past one," *Ghana Armed Forces News*' Comfort Akwagyiram observed in 1976, "few wives are engaged in this 'Abongo wife's role.'"[148] Instead, wives' organizations organized sewing, cooking, and literacy classes.[149] Prominently beneath the "2nd Infantry Battalion in Action" headline, *Ghana Armed Forces News* ran the subtitle: "The Women at Home and the Men on the Beat" in the Sinai.[150] Beneath a photo of Ghanaian women studying dutifully at school desks, a caption read, "I would not be surprised if you find your wives writing your letters on your return home."[151] Beneath one of a dozen women cooking: "How happy would the husbands become if they return home to discover that their wives have acquired skills to prepare various dishes." To Lydia Baafi, wife of the commander of the then-current Ghanaian UNEF II detachment and chairman of the "Wives' Club" in 1976, life for army wives whose husbands had deployed was considerably busier than usual. She had organized "basketball, netball, and athletics" for the other wives but also "classes in cookery, beadwork, needlework, [and] tie and dye."[152] "We try as much as possible to depend on some of the wives who are experts in their various fields," Baafi told Akwagyiram. "For instance, the nurses among us once in a while organize lectures on child-care."[153]

Peacekeepers' wives used the deployments to access formal education and build supporting relationships with other families, and former domestic roles changed as women became more independent, although they did not change much. Even while celebrating the possibility that an army wife might become an excellent, literate chef, *Ghana Armed Forces News* suggests that these new skills primarily made her a more supportive wife—able to please her husband with love letters when he was away and with dinner when he returned. By promoting these small educational initiatives for wives of deployed soldiers and distinguishing them from the typical "Abongo wife's role" of communal cooking, sweeping, and waiting for pay, the Ghana Armed Forces' official news magazine equated deployment with an occasion for wives' transformation.[154] They became more educated, more independent, and less connected to regional or ethnic identities. Becoming a peacekeeper's wife was another step beyond the "army wife" on a continuum that ended with modernity.

Some of the same colonial tactics of reshaping African women's expectations for work, family, and property in order to modify the (male) African labor force into modern citizens continued in the Ghana Armed Forces even after independence.[155] Especially in British colonial armies, processes for creating reliable colonial soldiers mirrored those that colonial powers pioneered for creating industrial citizens.[156] Hygiene inspections, home inspections, and officially registering wives were all part of the European program of "detribalization."[157] Around British colonial army posts, young African women adopted the clothing, hairstyles, perfumes, and cosmetics linked to Western conceptions of "femininity."[158] Even after independence, Ghanaian soldiers required their commanding officer's permission to marry; Henry Anyidoho recalled his pride after his commander complimented Mercy as "a good choice."[159] Ghana Military Academy closely scrutinized cadets' courtship behavior in the 1960s and 1970s. "As a gentleman officer-to-be, one was expected to know how to handle the lady partner," Steve Oduro-Kwarteng remembered.[160] "The officer cadet was watched as to how he handled or treated his partner," and "we were told that it was the pride of any unit commander to have his young officer call on him to introduce his wife as the lady who had been his girlfriend while he was at GMA [Ghana Military Academy]."[161] The Ghana Army regulated enlisted soldiers' marriages as well, reminding service members in 1975, for example, that they should register all their wives with the service to permit them to enter and live in the barracks—even though only one wife was entitled "free medical care in military Hospitals."[162]

In the era after independence, however, Ghanaians—both elites and ordinary citizens—also seized the rhetoric of modernity to made demands against their own states and against the international community. For Africans, Frederick Cooper argues, modernity had concrete meanings, including "health facilities, education, decent pensions, opportunity to sell one's crops and obtain useful commodities from elsewhere."[163] For them, modernity was a policy as much as a theory, and the language of modernization gave them a basis for asserting claims: "if you think we should be modern, help us find the means."[164] Criteria for what constituted a modern, successful home reached Ghanaian soldiers and their wives through extensive international travel on courses abroad—like the Informational Program—and through regular cycles of UN service alongside military members from around the world.[165] *Ghana Armed Forces News* also reinforced those standards, for example, celebrating the "international army wife," encouraging wives to take reading or cooking lessons, or reprinting the chief of staff's exhortation to deploying soldiers to invest their money carefully in their "respective ventures."

Ghanaian military families seized the prescriptive rhetoric of the modern military to assert such claims for pay, benefits, and housing. In order for a Ghanaian woman to become the "international" army wife that *Ghana Armed Forces News* praised—that is, an independent mother living in isolation from her extended family—she required specific tools: a sewing machine to prepare and mend garments by herself, and from home; a refrigerator to enable her to live farther away from the community and shop less often. UN service helped purchase tape players, radios, and television sets from abroad that joined military families to new and less communal modes of domestic entertainment. Ghanaian soldiers abroad carefully saved their money to apply their family strategies, acquired paving machines and roof tiles, for example, to accommodate new standards for the nuclear family living in detached quarters. Training courses and UN service abroad provided the opportunity to demonstrate new norms for domestic wealth and insulate soldiers from capricious national military budgets. Through the homes they built, the household appliances they acquired, and the new, mobile identities military service celebrated, Ghanaian military families both labored under external standards of modernity and shouldered them. New modes of living around army posts demanded new household appliances, but Ghanaian men and women also cooperated to leverage their international military service into acquiring those new markers of domestic success.

The small changes in education and outlook also reinforced the women's disconnection with former domestic roles, especially given the genuine difficulty of separation. Unfortunately, only a few bits of historical evidence recording peacekeepers' wives' own words on that subject remain. In September 1974, describing a recent picnic for the families of deployed soldiers, *Ghana Armed Forces News* quoted several of their letters:[166]

"Almighty God," wrote the Battalion *Magazia* Abiba Belu, "give Lt-Col Nyame [the detachment commander] patience, sympathy and encouragement. . . . We know the life is not an easy one there, but with your dynamism and fortitude all will be fine. If they keep on drinking because of the difficulties, please convince them to stop." To her husband, she wrote: "We pray always for your well-being. Live in peace and work harder for promotion."

Abla Gadzo Abordo reassured her husband Raphael, "My pregnancy is not troubling me. Pray for me to deliver a bouncing baby boy for you to become a soldier like you."

Esi Boridziwor Okine gave her husband wonderful news: "I've given birth to a beautiful baby girl for you this month. She has taken your pointed nose."

Stephen Menu also had good news for his father, Akwai, "Your big cat has delivered three kittens and the rabbit too has had five."

Ofori Mensa told his dad, "Papa, I pray that nothing will happen on you at that place."

Not all wives or children were so enthusiastic. "That day you left me, it made me sick. I couldn't eat the next day," wrote Beatrice Mamle Cudjoe to her husband. "David, at times I feel so lonely that I find it difficult to sleep."

"I feel miserable without you," wrote Faustina Amegbletor to her husband, John.

"Don't let anybody deceive you to do bad things," wrote Cicilia Arkason to her husband, Michael.

"We want to read from you," Mary Boatemaa begged her father, Micheal Atta. "People are showing off with letters from their husbands."

Finally, Deborah Ketor reassured her husband, John, "When you are worried sing 'Across the Bridge' by Jim Reeves."

"Comrades," a sergeant from the Rear Party reassured them, "all is well and groovy. Your wives are always happy. Don't be disturbed at all. Work hard to come back as worthy ambassadors."[167]

These family reflections were both typical and unique. Private David Gamor, correspondent for *Ghana Armed Forces News,* supposed that "these feelings expressed by the wives and daughters of the soldiers serving in the Middle-East could come from anybody at anytime who finds himself in the same situation."[168] Every family similarly separated in the diaspora would recognize their sorrow, heartache, and fear. In this case, though, specific circumstances of international military service distinguished this separation from others. They prepared and shared their letters at an army-sponsored picnic; even at their most intimate, the messages passed through army hands with army approval. Wives prayed for their husbands' safety, but also that they would serve honorably within the army in order to merit promotion. They lamented their separation, but some still prayed for their sons to become soldiers as well—as if willing their own separation onto future generations. On the other hand, identifying Jim Reeves as the preferred method of consoling a lonely husband is as good an indication of a transnational identity as any. Even the article's title—"Sentimental Journey"—evoked Ghana's heroic participation in Burma during World War II.[169] Wives recognized that their husbands and fathers were in danger, but they also acknowledged that their families' sacrifices were, on some level, connected to a larger project of international peace.[170] Unique political and economic conditions characterize all diasporic communities; in this case, Ghana's service with the United Nations shaped this one.

* * *

Continually since 1973, hundreds of Ghanaian soldiers deployed to the Middle East on seven-month peacekeeping missions. This section briefly explored the national and corporate incentives that convinced Ghanaian military leaders to pursue and sustain these missions in order to contrast the ways individual service members and their families understood their deployments. UN reimbursements provided millions of dollars to Ghana's struggling economy in the 1970s and 1980s and subsidized armed forces' training and equipment beyond what the Ghanaian government could afford. Serving alongside soldiers from around the world made Ghanaian soldiers "more aware of international norms for the treatment of soldiers," and their meritorious service became a source of national pride.[171] For Ghanaian families, UN service offered pay and material rewards that few other Ghanaians could access during that time. Soldiers' families built houses, furnished their homes with refrigerators and radios, started businesses, and supplemented their income selling scarce household goods and appliances. International military service imposed new domestic standards on soldiers' families, but soldiers' families also employed that service to fashion their own responses. But these economic opportunities came at a cost: lengthy separations.

Ghanaian soldiers' wives' associations mitigated their sense of isolation with programs building both personal independence and solidarity with other wives. They often displayed variously local and international identities that connected military families to one another regardless of regional or ethnic background, privileging global norms of military service that transcended national identities. They expressed their loneliness and anxiety from their separations in typical ways, but the specific reason for their separation was very different from other cases of labor migration in the post-independence African diaspora. It is to that mission, its danger, and Ghanaian families' understanding of the emotional significance of being "peacemakers" that we turn next.

## Writing for Peace

Both leaders and ordinary members of the Ghana Armed Forces used combinations of local, national, and international reference points to explain their decision to join the army and evaluate the benefits of military service for their families' lives. Since independence, Ghana's government and Ghanaian military leaders aggressively pursued education and training opportunities abroad, partly because Ghana lacked the capacity to provide advanced training on its own and partly to maximize the amount of military assistance it received. Service members themselves also roundly appreciated the experience of international education. When preparing his massive history of the Ghana Armed Forces, Stephen Kojo Addae recalled, "This author has been told again and again, by retired officers trained in the UK, the salutary experience of training with men from all over the world."[172] After retiring from the army as a major general, Henry Anyidoho believed, "one of the greatest benefits I got from the Army was the acquisition of knowledge. I received training from all sorts of institutions, locally and overseas."[173] International education was common for Ghanaian soldiers, especially officers, and they routinely praised the experience as educational, inspirational, and enjoyable. It sometimes offered safety from both physical and economic danger as well.

Compared to international education, UN service offered similar financial incentives but was more dangerous and affected far more Ghanaians. As many Ghanaian soldiers served as peacekeepers in UNIFIL in one six-month rotation in 1983 as the number of all International Military Education and Training–sponsored students at American training facilities in thirty years between 1962 and 1992.[174] Service with the United Nations offered neither comfort nor safety—six Ghanaian soldiers died in the Sinai between 1973 and 1979; twenty-two died in Lebanon between 1979 and 2000.[175] In their descriptions of their UN service, however, Ghanaians expressed solidarity with a global military ethos that surpassed Ghanaian regional differences, celebrated Ghana's global leadership in the international order the United Nations represented, reinforced Ghanaian pan-Africanism, and offered psychic rewards to individuals as "peacemakers" that transcended ethnic, national, or pan-African identities.[176] Just as Ghanaian students had when on courses abroad, individual Ghanaians serving with the United Nations expressed a form of transnational solidarity only accessible through service in a global military endeavor.

* * *

As Kwame Nkrumah had in the early 1960s, Ghanaian political elites in the 1970s and 1980s—civilian and military—considered the United Nations essential for Africa's political and economic improvement, Ghana's included. Participating as equals in the UN General Assembly empowered all newly decolonized states, Ghanaians above all, even when Cold War powers attempted to thwart their collective will.[177] African nationalism "was of little consequence" to the United Nations in 1945, Kwame Nkrumah wrote in 1970. "Since then, however, so many former colonies have achieved independence that Afro-Asian countries now form the most influential single group within the United Nations."[178] William Ofori-Atta, Ghana's minister of education, reminded participants at the African Regional Symposium on Disarmament and Development in July 1970, "The new nations of Africa and Asia are the most ardent supporters of the United Nations and also the strongest advocates of a world order."[179] Ghana secretary for trade Kofi Djin expressed the same belief at a flag-raising ceremony in Accra celebrating the United Nations' fortieth anniversary in 1985. "There could be no alternative to the UN," he said, "to which small nations like Ghana would continue to look [and] use as a forum to discuss and exchange views on issues that affect the global peace."[180]

Ghana's ambassadors acceded with delight to Secretary-General Kurt Waldheim's requests to reaffirm Ghana's participation in each UNEF and UNIFIL mandate as the Security Council renewed them—roughly every six to ten months.[181] Ghana's government accepted the secretary-general's request to increase its UNEF detachment in 1975 and again in 1976 when Brian Urquhart, undersecretary general for special political affairs, struggled to find another African state to replace the withdrawing Senegalese, Nepalese, and Panamanian contingents.[182] The secretary-general requested that Ghana maintain its full contingent of 595 even when the rest of the UNEF II force shrunk from 4,895 to 4,215 in 1977.[183] He requested an additional 180 Ghanaians to join UNIFIL in Lebanon in 1983, and Ghana's contingent there grew from 400 to nearly 600.[184] As with all other requests to extend the UN mission's mandate, Ghana's ambassador pledged, "Ghana would continue its contribution for as long as called for."[185]

Officially, Ghanaians considered their participation in UN peacekeeping a great global compliment, and they imagined vast benefits to flow from it. "It is common knowledge that in any belligerency, especially between two people or nations, one needs a neutral person or nation to stand in between the two parties in order to bring peace," *Ghana Armed Forces News* reported in 1976.[186] "It is needless to say that such a person or nation need be peace-loving. Thus when the Ghana Armed Forces and, for that matter Ghana, was invited to take part in the peacekeeping operations in the Middle East in 1974, the obvious was implied." "I think the whole Middle East problem rests on the shoulders of the Ghana contingent," said the contingent commander, Lieutenant Colonel Kwame Baafi, in 1976.[187] Victor Ansah, correspondent for *Ghana Armed Forces News*, argued that by helping reopen the Suez Canal, Ghana served "world humanity and civilization" and benefitted Ghana and the entire world by helping lower oil prices.[188]

*Ghana Armed Forces News* correspondents and contributors to the United Nation's peacekeeping journals emphasized Ghanaians' equal participation in the peacekeeping "melting pot." Captain John Amissah wrote in *Litani*, UNIFIL's monthly news bulletin, "The different members of the organization, with their varied sociocultural backgrounds, entails regarding the UN, and for that matter its peacekeeping offspring UNIFIL, as a 'melting pot'—a common ground for the blending of ideas and cultures for the salvation of mankind."[189] "No individuals or nations," he warned, "should do anything to derail the UN ideal, otherwise they would be adding more fuel to the fire to explode the 'globe.'"[190] Military police from Fiji, Finland, France, Ghana, Ireland, and Norway might "vary in their religious beliefs, language, race, culture, and even in their military training and equipment," Amissah wrote from Lebanon, but with "mutual understanding and cooperation," all desired "fostering greater brotherhood among the different nationals."[191] Lieutenant Makana Blay recalled the advice his commanding officer—"that good old 'Osofo'"—gave as he departed for the Sinai in 1975: "We should be good ambassadors whilst in the Middle East because we are 'citizens of no mean country.'"[192] "Our men can talk face-to-face with other nationals without any feeling of inferiority complex," said Blay. "They know that they are and can be as efficient as any other soldier from any part of the world."[193] *Ghana Armed Forces News* reprinted internationally obtained compliments with pride. From UNEF Chief of Staff and Indonesian Brigadier General Rais Abin: "I am profoundly impressed with the Ghanaian soldiers."[194] From a

Canadian colleague: the Ghanaian soldier's "ready smile is inspiring for all of us and very good for high morale."[195] A Nigerian diplomat: "Our brothers serving with UNEF have a high sense of dedication to duty, but above all they keep their national consciousness and work hard to put Ghana on the world map."[196] For the editors of *Ghana Armed Forces News*, equal status among deployed forces reinforced Ghanaian nationalism.

Peacekeeping seemed to cement the national identity of Ghanaian soldiers, or at least the Ghanaian government attempted to subsume regional or ethnic differences into a larger, unified Ghanaian culture and present it for international consumption. At the Medal Day ceremony in 1985, as one detachment of Ghanaians received their UN service medals and prepared to return to Ghana, the contingent staged an elaborate demonstration of music, clothing, dance, and ceremony by small groups claiming to represent Ghana's largest ethnic subdivisions. "Like other ethnic groups of Ghana, the Ewes have reverence for their ancestors and cultural values," Stephen Owusu wrote in *Litani*.[197] A group of Ewe, "adorned in various paraphernalia" and "depicting a chief and his elders," demonstrated a typical "festival scene" in an Ewe kingdom. Next, a group of Ga men "offered another rich aspect of Ghanaian culture in the form of the Kpanlogo dance which was so popular among Ghanaian youth in the late '70s."[198] Finally, a soldier dressed as an Akan chief. "He wears a headgear and sandals plaited in solid gold. The gold trinkets around his neck and wrists weigh the hand so much that the hand has to be supported by a servant before a handshake."[199] In photos, the chief and his retinue, arrayed in Kente cloth and gold, "dance[s] to the 'Adowa' tune," while a "'Fetish Priest' evoking his spirits" purifies the ceremony.[200] To Owusu, "This is how Ghanbatt presented its country to guests at the Medal Day celebration."[201] This was also how they reaffirmed the idea of Ghana to the diverse Ghanaian unit itself.

In stories like this, Ghanaian identity appears for fleeting moments, expresses itself, and then recedes behind other international, global, or individual ones. At a different ceremony three months earlier where the UNIFIL force commander issued UN peace medals to the Ghanaian and Irish battalions, the Irish Catholic chaplain and Ghana's Muslim imam prayed side by side—one in English, the other in Arabic—for "those connected with the task of working for peace."[202] After parade formalities, both groups migrated to the Mess for "a few minutes of throat wetting," a "jaw-breaking concert by Irish soldiers which sent a stab of pleasure into the hearts of the audience," and a Ghanbatt band concert "with a blend of classical, Ghanaian highlife, and reggae tunes." Within just a few hours, Ghanaian soldiers expressed global religious identities (Christian and Muslim prayers), solidarity with the United Nations and the internationalist peacekeeping ethos that united small, troop-contributing states like Ireland and Ghana, and a moment of pride in the national band's national music. Finally, as if forgetting it all "for that short evening, the men danced away the difficulties in the task of peacekeeping."[203]

In the way it boosted Ghanaians' pride while simultaneously shredding national distinctions in transnational social events, the Ghana contingent's band epitomized peacekeepers' churning national, international, and global identities. The Ghana Army's Sixth Infantry Battalion ceremonial band joined the first Ghanaians who arrived in the Sinai in 1974. They played martial music for official purposes, but at night they became "The Ranchis," playing dance music gigs for peacekeepers around the Sinai.[204] "For those who may raise an eyebrow to ask about the sanity behind sending bandsmen to the war zone, I will like to refer them to the Vietnam War when the US government provided its troops with the James Brown show band," public relations officer Captain Victor Ansah wrote, with international aplomb.[205] The Ranchis' successors, the HOT BARRELS became local celebrities after Ghana's contingent commander loaned them to the Swedish contingent for a night.[206] "We were worried as to what type of music would appeal to the Swedish," Lance Corporal Raphael Tay remembered.[207] "But we got over it" when the band was "mobbed after doing the African beat 'Shakara.' The shouts of 'Da capo,' (enchore) were enough to assure the bandsmen that they were prepared to take in any number Western, African, Latin, Congo etc., and the band dug it out to them. That was a night to be remembered. A show that was scheduled to last two hours from 8 p.m. local time, ended at 12 p.m."[208]

The HOT BARRELS were Ghana, Africa, the United Nations, and the globe all in one. At a gig at UN Headquarters in Ismalia, Egypt, they mixed American ballads like "Killing Me Softly" with Ghanaian highlife and their own compositions "Buffer Bone" and "To Sinai and Back." "No less than 15 different countries [were] represented on the staff of the UN Headquarters," and the HOT BARRELS kept them all spellbound, proving to "the multi-racial set up what Ghana could offer aside its military duties." After the HOT BARRELS jammed for African ambassadors to the Arab Republic of Egypt at the African Diplomatic Club in Cairo, Ghana's ambassador gushed, "You guys have made me proud among my friends."[209] The band was "the best in Egypt," and "signed a month agreement with a popular night club and hotel 'La Ronde.'" "The 'HOT BARRELS' and the Ghana contingent has made Ghana known," Ansah crowed. "Very soon, we shall not be called 'Ghana Bissau' by our brothers, the Egyptians."[210] It is not clear if "Ghana Bissau" was an Egyptian slur against black Africans in general or just against Ghanaian provincialism, but Ansah boldly seized "brotherhood" status from the Egyptians nonetheless. Plus, "armed with a few pamphlets on tourism in Ghana" he added, the band also promoted tourism, "a foreign exchange earner"—a real concern in the deepest days of Ghana's economic crisis and a recurring theme in *Ghana Armed Forces News* as it struggled to justify military spending in Ghana. The HOT BARRELS returned to Ghana in February 1975. Their replacement, "Third Generation," apparently "played so well that it made the music of the Hot Barrels taste sour in the mouth of those who had swallowed mouthfuls of it."[211] Each deployed unit brought a band, and the band brought part of Ghana with it. But at gigs for multinational audiences, the band also transcended Ghana, and men from around the world danced.

Finally, Ghanaian soldiers expressed solidarity with an international military identity in the 1970s and 1980s in the ways they yearned for peace. In one of the very first dispatches from the Sinai, before Medals Day parades became common, before the massive reimbursements to Ghana's soldiers and government, before anyone knew that Ghana would continuously maintain five hundred to seven hundred soldiers in the Middle East for the next four decades, public relations officer A. A. Enninful reflected: "It may be a difficult task, it may be a thankless endeavor and surely their reward shall not be in gold or silver. . . . Beyond this, there is definitely a richer something—the Beatitudes' reward for the peacemakers: 'they shall be called the children of God.'"[212] Blay's "Fanara Diary" mourned for Egypt's children: "Poor innocent children! Some of them lost their fathers during the war. How I wish there is peace so that huge sums of money being spent on arms would be used on their upbringing."[213] Warrant Officer Stephen Owusu, who served three tours in Lebanon in the 1980s as a correspondent for the Public Information Office, collected and republished his most poignant articles in 2009 in *Writing for Peace*, taking the title from a *Litani* article he wrote in 1985.[214] "Before my maiden physical contact with Lebanon," he wrote, "I had only a faint idea of the country. . . . I remember one particularly ghastly scene in a TV news slot, where street battle in Beirut left the streets filled with horror, reminiscent of a battlefield. My reaction to what I saw was that of shock and sympathy for the defenseless civilian population (made up of children, women and the aged) I saw fleeing in panic for dear life."[215] After that, he recalled, "I nursed the urge for an opportunity to help in my little way the efforts of the United Nations to restore peace in Lebanon."[216] In 1985, he returned to Ghana after his third tour in Lebanon "with my hands raised in prayer for the country I love and have sacrificed so much for to see the light at the end of the tunnel, for my sacrifice and that of my United Nations colleagues will be futile if it is not linked up with total peace for Lebanon."[217]

Ghanaian soldiers expressed their longing for peace in *Ghana Armed Forces News* and in the "Poet's Corner" of Ghana's peacekeeping magazines. In *The Marakah*, Ghana's journal from Lebanon in the late 1980s, Warrant Officer Sowa Okpoti wrote:

> Allow me to do my duty to God and man
> To a people famishing for peace and security in this turmoil
> The restoration of dignified life in tranquility is my lot
> Though I'm but a spark in the peace process
> Surely, I've something up my sleeve to surprise the world with.[218]

In another poem, "Mediterranean Sea," Okpoti wrote: "Mother Mediterranean here we are, we beseech / Great Demarcator of two terrains, here we are, we kowtow . . . Peace and security being our search / Your sympathy and empathy we crave."[219] In 1990, *Ghana Armed Forces News* published Private Ahwireng E. D. Atore's poem "Peace":

> I'm Peace
> I'm Odomonkoma (God)
> I love peace so I created it . . . I warn you!
> Keep peace and prevent destruction
> Birds which run north, south,
> east and west even announce
> my message of peace
> I pray thee
> Keep peace.[220]

For these soldiers, Ghana's military publications were perfectly acceptable fora to express a global desire for peace. More than that, the soldiers considered themselves global citizens—as sons of the "Mother Mediterranean"—who actually *brought* peace to the world. Their sentiments were not simply those of longing for peace in their troubled home; they expressed their passion for the project of peace that they had undertaken in another place, alongside comrades from other places. For Owusu, "the country I love" was Lebanon; Okpoti considered himself "but a spark in the peace process" but one with "something up my sleeve to surprise the world." They desired peace, and they genuinely considered themselves part of the international structure that delivered it. They could only do this from outside Ghana, but their decades of prior experience integrating international education, international intellectual and political influences, international travel, and international economic strategies made soldiers from the Ghana Armed Forces adept at expressing their international identities on peacekeeping missions abroad. Many of the same influences made their families equally adept at manifesting multiple international identities as well.

## Conclusion

This chapter uses examples from Ghanaian men and women who participated in a massive global network of military education and training and peacekeeping operations to answer the question: "How did ordinary Ghanaians' lives change as a result of military-sponsored international travel?" It proposes a new interpretation for modernity that combined individuals' alternate political and economic strategies with a global ethos of peacemaking. That identity intersected with the internationalism that other military travelers and communities around global sites of international military exchange also expressed.

Ghanaian families found in international education and international peacekeeping service a combination of resources for personal protection, both political and economic, as well as new demands and new opportunities for a new outlook. Many Ghanaian families chose military service after independence specifically because it offered the best opportunity to participate in a social imaginary that celebrated "going abroad to get ahead." Given Ghana's political uncertainty between the 1960s and 1980s, Ghanaian military families abroad simultaneously depended on the Ghanaian state for transportation to the sites of international contact and were unable to rely on the state to continue to support them while abroad. Some, such as the Anyidoho family, the Oduro-Kwartengs, and the Erskines, even employed international military agreements specifically to protect them from their own nation-state. The Ghanaian state, for its part, similarly leveraged its participation in international peacekeeping missions to alleviate its own economic crises by claiming millions of dollars in reimbursement from the United Nations for its troops and equipment in Egypt, then Lebanon. But even when questioning the long-term benefit of providing peacekeepers as "Askaris of the New World Order," more than simple "mercenarism" motivated states like Ghana to participate in UN missions.[221]

Even larger numbers of Ghanaian soldiers gained access to international travel, service alongside soldiers from around the world, and generous reimbursement from the international community performing peacekeeping for the United Nations. Thousands of Ghanaians—up to 31,000 people, counting repeats—deployed to the Sinai and Lebanon between 1973 and 1992. While deployed, they acquired household goods such as appliances, furniture, and building materials that permitted them to execute personal economic strategies that both insulated them from further dependence on the Ghanaian state and resembled economic strategies of Ghanaians in diaspora communities elsewhere around the globe. One essential difference, however, was the way military wives adapted to the isolation resulting from lengthy separations and the danger of military service, even peacekeeping service. They responded by learning new skills and forming new communities, with the paradoxical result that they conformed more closely to patriarchal ideals about the "ideal military wife" while at the same time acquiring the resources to become more independent. They developed an international identity around national military service that required mobility and prescribed internationally defined standards for modernity while simultaneously offering resources only available from international sources.

International education and peacekeeping service did more than fatten Ghanaian soldiers' pockets, however. In the ways they expressed solidarity with other military families abroad—especially around institutes for education and training—they manifested the same internationalist ethos found in communities of military families elsewhere. International military travel cultivated personal relationships, and an international identity resulted. Peacekeepers, likewise, expressed through reflective essays, poetry, and their own memoirs their commitment to global peace after serving with the United Nations in the Middle East. Such service clearly reinforced their national pride, and the Ghana Armed Forces defended its size, budget, and corporate benefits based on this national pride, but individual Ghanaians' pride also transcended strictly national boundaries.[222]

# 4 "I think they would be better off if we took them on a tour of Disney World"

## *Continuity and Change in International Military Education and Training*

### Momentum or Inertia?

On Friday, June 22, 1990, in a conference room in the Dirksen Senate Office Building where three witnesses sat before three senators, Senator Patrick Leahy (D-VT) posed triumphantly beside a small stack of papers—a General Accounting Office (GAO) report of thirty-five pages. "I understand none of you has had the chance to read it," he boasted modestly, "so let me just read part of the findings."[1] Indeed not; the report did not exist a week before and Leahy specifically requested it be sequestered until the twenty-fourth.[2] He knew it so well because he had personally commissioned this GAO study a year prior and had received a private summary briefing a month earlier, confirming what he had thought for years: "IMET [International Military Education and Training] is a program that runs on momentum or inertia," he declared. "Until this GAO study I requested, I do not think that anybody had looked at IMET since the early 1970s." He was partly right.[3] The conditions the report exposed, including insufficient measurement and the tendency for the United States' undemocratic allies to eagerly accept US military education and training grant funds, had persisted since the early 1970s. In fact, they had persisted far longer than that.

Despite Leahy and his GAO report's long-festering criticism of the program that funded education and training for foreign military personnel in the United States, both Leahy and the GAO continued to endorse the program because neither doubted the ideology behind it. The GAO accepted that international military education and training programs successfully exposed participants to "US values, citizens, institutions, and commitment to human rights."[4] They both cited a February 1990 study by the Defense Security Assistance Agency (DSAA) as evidence, repeating the premise that "the

IMET Program is one of the most cost-effective programs for pursuing US foreign policy and national security objectives."[5] As he questioned Lieutenant General Charles Brown of the Defense Security Assistance Agency later in the hearings, Leahy quoted that GAO report, which in turn cited a DSAA article that was itself an expanded version of the testimony Brown had given to Leahy's Senate Appropriations Committee one year earlier.[6]

Brown stated in 1989 and repeated in 1990: "IMET has been generally recognized as our *most cost-effective* foreign assistance program. IMET is a people program that establishes valuable personal relationships and lines of communication with foreign military personnel, many of whom rise to prominent positions."[7] "Foreign military students learn our ways and are exposed to our values of support for democracy and personal integrity based on pride in military professionalism," Brown said. "They, in turn, help spread these values in their countries by sharing their experiences with military and civilian counterparts."[8] That revolving door of congressional evidence also exemplified the persistent assumption—that international military training produced psychological transformations—that had justified the program for three decades.

Leahy and the GAO's candid critique of IMET's "inertia" and its inability to measure its own progress were only the last in a series of periodic attempts throughout the 1970s and 1980s by Congress, the Department of State, and the Department of Defense to evaluate the IMET program. This chapter shows the results of those attempts and argues that despite occasionally pointed criticism over that period, few deviated from the long-held faith in international military training's potential to transform visiting military officers psychologically—at a low price. The case of Ghana exemplifies the resilience and plasticity of that faith, as American defense planners gave little thought to the rise of American-trained Colonel Ignatius Acheampong in 1972 and even less when interpreting the meaning of his 1977 overthrow and his execution in 1979.

As the 1960s closed, academic and diplomatic justifications for American military assistance policies evolved away from modernization theorists' desire to transform the "developing world" and embraced the more decentralized and frugal Nixon Doctrine. Political intervention by African, Asian, and Latin American militaries seemed the norm in the early 1970s, and the US government struggled to rationalize its military assistance policy against ongoing military incursions against democracy and civil rights. Détente with the Soviet Union and China made the Nixon Doctrine—which expected the United States' allies to assume greater responsibility for their own security—look more like a tactical retreat than a total rout.[9] In 1970, Nixon's special

Peterson Committee—named after the Bank of America president who chaired it—encouraged more private investment overseas, recommended that multilateral institutions such as the World Bank replace unilateral American economic assistance, and exalted the power of "market forces."[10] "If the goal is economic development," the committee reported, "the issue is one of efficiency, not ideology."[11]

Peterson also re-emphasized the long-held official justification that international military education and training made foreign militaries "self-reliant."[12] Meanwhile, the United States continued to support "bureaucratic-authoritarian" states like Brazil, Argentina, and South Korea and "Third World policemen states," such as Turkey, South Africa, Iran, and Indonesia, with military assistance and training, while Washington interfered as little as possible in their democratically suspect domestic politics.[13] The "New Approach to Foreign Aid" no longer considered massive physical, economic, or ideological intervention in most of the Global South essential for US security, but international military education and training programs remained an inexpensive way to preserve American diplomatic initiative, just in case.

Times changed. Even before Jimmy Carter's election, "internationally recognized human rights" became a major concern in American assistance policy.[14] To the extent that it tolerated other powers' spheres of influence, détente was incompatible with the moral universalism inherent in American traditions of liberalism.[15] Meanwhile, in response to austerity in the early 1970s and against challenges to military assistance for regimes that violated their subjects' human rights, the State Department, Department of Defense, and Congress attempted to evaluate the effectiveness of US aid programs. When it came to programs to sponsor foreign military training in the United States, they repeated old justifications, invented new ones, and ensured that the scope and purpose of these programs changed little for the next twenty years.

Numerous and persistent military regimes in Africa, Asia, and Latin America challenged American diplomacy. Some scholars argued—and policy makers agreed—that military rule impeded national development in the Global South, without admitting that US military aid was responsible for the prevalence or persistence of military rule in the states the United States' supported.[16] Others noted that US allies were often highly modern, or possessed high organizational strength, but fell far short of democracy.[17] When evaluating American military assistance programs, however, the State and Defense Departments maintained their faith that programs that demonstrated in person the American model of civil-military relations had the most promise to undo the trend toward military rule. Curiously, for example, Ghana's 1972 coup by a US-trained army officer prompted little self-reflection in US policy makers. Even scholars who lamented the persistence of military rule noted only tangentially that Ghana's new military head of state had graduated from Fort Leavenworth's Command and General Staff College.[18] To some extent, this apathy was a symptom of a general decline in interest in West Africa among US foreign policy makers in the 1970s.[19] But State Department officials and scholars also implicitly accepted what applied to many other newly decolonized states as well: Ghana's domestic political situation had largely Ghanaian explanations, and in Ghanaian foreign policy, US aid competed with other immediate issues like resolving Ghana's international debt and ending apartheid in South Africa.[20]

In the mid-1970s, in response to congressional, then presidential, concern over the United States' image abroad and its responsibility to defend "internationally-recognized human rights," the training program quickly adapted and adopted the new human rights paradigm. Despite scant evidence, its administrators continued to boast of its long-term success and continued low cost. These criteria enabled the IMET program to weather ideological and budgetary scrutiny in the late 1970s and throughout the 1980s, often recycling year after year the same justifications that had worked for decades.

The persistence of military rule in Ghana never seriously threatened American justifications for its ongoing military training and education grant program there. In the 1970s, scholars and policy makers clashed over what caused military rule and whether American influence perpetuated it, but neither side expected military rule to go away.[21] Throughout the period, and despite scant supporting evidence, Defense and State Department planners continued to recommend international military education and training programs as an ideal method of eliminating military rule from the Global South.[22] They did so partly because the specific international military education and training program was relatively inexpensive, it conformed to new trends in American foreign policy de-emphasizing direct intervention, and later proved compliant to congressional attempts to wend their concerns with human rights into American military assistance policy.[23] IMET was the United States' only program for military assistance to Ghana, but when its anticipated outcomes failed to materialize, Defense and State Department policy makers re-emphasized the intangible, long-term, and unaccountably cost-effective benefits of international military education that had made modernization theorists embrace it years earlier.

Often, though, American military assistance planners exploited their old faith to offer new rationalizations for international education. Annual justifications for the United States' small IMET grants to Ghana dynamically adapted to rapid political change in Ghana after 1979. IMET promised first to uphold the ideal of military professionalism in Ghana, then to temper the radicalism of military-led revolution in Ghana, then to rebuild the strained diplomatic relationship between Ghana and the United States that followed Rawlings's rise to power. Without ever changing the underlying faith in the power of international military education and training to transform individuals, American planners continually revised their estimates for the corporate, political, and diplomatic benefits that IMET promised to obtain in Ghana.

Either through "inertia," or "momentum," between the 1950s and the 1980s, the same fundamental assumption that made foreign military training a natural instrument for modernization theorists persisted: life among American communities yielded lasting psychological benefits for foreign visitors that over time transformed their home states and reinforced American influence there. International military training appeared to be sound foreign policy. Best of all, compared to the rest of American foreign aid, this "people-to-people" program was always a bargain.

## The Informational Program in 1971: Inertia, Ideology, and Internationalism

Neither Congress, the State Department, nor the Department of Defense resolved the philosophical and jurisdictional conflicts over who was best equipped to formulate, execute, and supervise American assistance programs. By 1972, officials in the State Department struggled to reassert their relevance and image, lamenting, "Congress tends to approach executive branch programs with suspicion, intensified by the impression of vast, essentially superfluous numbers of personnel and lack of leadership within the bureaucracy."[24] Congress and the American public neither cared nor supported security assistance ("just a euphemism for foreign aid"), and they considered the State Department "particularly ineffectual."[25] Neither the Peterson Committee, Nixon's decentralized foreign policy doctrine, nor internal attempts to validate and measure American military assistance programs resolved these conflicts in the early 1970s. In this turbulence, the simple assumptions that first commended international military education and training persevered, perhaps because of their simplicity.

* * *

By the mid-1970s, few scholars of Western and African relations approved of the prevalence of military rule in Africa, but this did not imply an academic consensus on how American military assistance specifically influenced or encouraged military rule. Huntington specifically argued that there while was no correlation between American military aid and military involvement in politics, there was also no evidence that professional training at schools like Fort Leavenworth had reduced the likelihood of military intervention either.[26] "Military aid and military training are by themselves politically sterile," he argued. "They neither encourage nor reduce the tendencies of military officers to play a political role." Princeton's Ernest Lefever agreed: "A disproportionate amount of attention has been given to conflicting claims about the indoctrination effect, good or bad, of US programs on the political attitudes of individual officers," he wrote, "when in fact, empirical evidence of [the Military Assistance Program's] impacts on individuals is sparse and generally ambiguous."[27] Before 1971, the Department of Defense and the State Department had not seriously attempted to find this evidence.

Proponents of military education and training programs accepted the lack of objective methods for measuring their effectiveness to justify their continuation throughout this period. In June 1971, the Nixon Doctrine's new emphasis on decentralizing American military assistance inspired G. Warren Nutter, the assistant secretary of defense for international security affairs, to request the chairman of the Joint Chiefs of Staff to review the effectiveness of military assistance programs that brought military elites from the poorer states to the United States for training and education.[28] Defense Secretary Melvin Laird tasked the State Department's inspector general of foreign assistance to perform an external review and solicit individual services' comments. Specifically, the inspector general's team examined the "Informational Program and Orientation Tours"—those portions of the military education and training programs explicitly intended to demonstrate those "significant facets of American life" to visiting military members at American army, navy, and air force facilities. The study's conclusions reiterated the psychological and political impact of foreign military trainees' visits to the United States but expressed frustration over the program's decentralized execution, inability to measure its effectiveness, and lack of follow-up with alumni of American military training and education programs.[29]

These objections were entirely predictable outcomes of evaluating a foreign policy program rationalized entirely by faith in universal attitudes and behaviors mythologized in transnational military camaraderie. They raised some practical objections, but inspectors Neil Campbell and Elliot Strauss from the State Department also reaffirmed the program's overall promise. Their report, and the reactions it prompted, reiterated the durable assumptions that shaped American military assistance programs through the rest of the 1970s. The 1971 evaluation dealt with international military education theoretically; an actual case of a graduate of American military education who overthrew Ghana's government in 1972 prompted very little reflection, and the evaluation offered no new basis to question the underlying faith, either in 1972 or for the rest of the decade. Quite the opposite, this faith expanded to incorporate the United States' newly concocted devotion to human rights.

As if undermining their effort from the start, Campbell and Strauss admitted in 1971 that there was "no yardstick by which the success of the [Informational Program] can be measured."[30] As both a symptom and cause of that problem, little data on graduates existed. When successfully located and interviewed, graduates of American programs were "apt to tell enquirers what [they] wished to hear." On one hand, the programs clearly created personal relationships and "a friendly feeling for the United States." On the other, the investigators noted, "there are a number of horror stories, Ali Sabry of Egypt, Qadhafi of Lybia [sic] and the President of Peru, all of whom profited by instruction at US military schools but can scarcely be termed friends of the United States."[31] Department of Defense officials even accused the two State Department researchers that wrote this report of getting it wrong. President Velasco of Peru received US military training before the Informational Program (IP) began, the Defense Security Agency's director argued.[32] And despite rumors swirling in the Arab world, Qadhafi never received military training from the United States.[33]

Whatever results flowed from IMET and related programs to serve foreign military visitors, these analysts admitted that they came cheaply. In 1972, 12 percent ($61.2 million) of the $500 million national budget for military assistance programs paid for international education and training for foreign visitors. The IP consumed just 2 percent ($1.37 million) of that segment, and just one-quarter of 1 percent of the entire military assistance budget.[34] Partly because they consumed so little investment in the first place and partly because they were the most visible American assistance programs to target states like Ghana, the programs avoided significant ideological scrutiny during the cycles of budget cutting of the late 1960s, 1970s, and 1980s. They did not remain above criticism, however.

The 1971 inspectors' report claimed that "virtually everybody" they contacted acknowledged that the program needed stronger central authority to govern the types of lessons and interactions the various schools' IP activities transmitted, but the report's specific recommendations belie that observation. The report's introductory cover letter even acknowledged that any improvements to the program must not destroy its "voluntary and spontaneous nature."[35] In fact, it was these characteristics—voluntary, spontaneous, and decentralized—that made the IP so representative of the nexus between official faith in the power of the average American to transform visiting elites and the average American's disdain for official academic and diplomatic niceties in favor of their own, local version of Americanism.

Sponsor families—those American families who assisted visiting military members by welcoming them into their communities, hosting them in their homes, or helping them navigate a difficulty with a base or community office—were the most useful. The inspectors admitted that the sponsor "usually becomes the [foreign military trainee's] guide, philosopher, and friend" with respect to the various aspects of the military post and US military life (11). Visits to American homes were the program's "most effective portion." "It is here that the trainee becomes disabused of some of his preconceptions of American life gained from the films and TV, and it is here that perhaps the strongest and most lasting friendships are formed," the report claimed (10). Trainees received "the most accurate view of life in America as it is actually lived" through these home visits (10). The inspectors noted that sponsors were "sometimes recruited through the initiative of the foreign training officer, sometimes by civic groups such as the Chamber of Commerce, the mayor's office, or volunteer women's groups" (10, 11). Helen and Robert Black of the Greater Kansas City Chapter of People-to-People International, for example, became Ghanaian major Ignatius Achel's sponsor family and remained in contact with his wife for years after.[36] Further, the inspectors acknowledged, American families had been receiving foreign military visitors into their homes long before the IP began in 1963 and "would probably continue if the program should disappear."[37] The official sponsor family program merely put a formal label on an attitude or collection of behaviors toward international military students that American families and communities, including "volunteer women's groups," already exhibited. Without saying so, Neil Campbell and Elliot Strauss acknowledged the "imagined world" of transnational connections and alternate paths toward international cooperation that undergirded American military education policy in general.[38]

The State Department's inspector general team published its evaluation of ninety-five Department of Defense–run military education and training programs in November 1971 and offered suggestions to become more efficient or effective.[39] They recommended, for example, that there be better centralization of different schools' programs, with a single Washington agency directing all programs; that the IP carefully "lessen the entertainment element" of its programs and tours; that staff levels for officers who administered the IP at various schools and bases be standardized across the services; that US agents abroad carefully screen each prospective student for

English language proficiency, since "if his proficiency is indeed so low that he cannot keep up, his frustrations may prevent the United States from winning a friend"; and that services keep better track of their foreign graduates' professional and personal progression.[40] The secretary of defense forwarded the State Department's findings to individual services and geographic commands in December 1971 and requested the services' responses to the State Department's findings.

In their responses, those agencies that were closest to the foreign officers jealously defended their freedom of action—they felt free to interpret the intent and application of the Nixon Doctrine according to their own initiative, and they considered the programs' decentralization necessary for their success. In other words, the officers who applied the American policy of training foreign military personnel and exposing them to "significant facets of American life" had their own conception of what made those facets significant and how best to pass their lessons along to visitors. They bristled at the suggestion that they needed further oversight from Washington.

Naturally, Defense Department agencies resisted criticism from the State Department inspectors, but they combined geopolitical savvy with a more internationalist sense of the benefits of human-to-human contacts to parry critiques. For example, the State Department report singled out for criticism an annual "deluxe treatment tour" of the United States for graduates of Brazil's and Argentina's national war colleges. Army general George V. Underwood, the commander of US Southern Command, who was responsible for all interaction between US and Latin American militaries, blasted the report's implication. First, he argued, the Argentines and Brazilians considered the tours "an integral, prestigious part of their study curricula" at their own war colleges and had specifically requested that the United States include all Brazilian and Argentine war college students in them.[41] In other words, the South American officers themselves considered a tour of American military, industrial, and government sites an essential aspect of integrating their own national military education with international content—especially if it was at the United States' expense—and this American general did not wish to disrupt that union of national and international purposes. Second, Underwood feared that eliminating these popular tours "could produce a host country reaction counter-productive to the intent of the Nixon Doctrine which advocates the maintenance of US influence and rapport with foreign military personnel, especially those who are in a position to influence national policies."[42] To him, decentralizing national diplomacy required increasing, not decreasing, the range of contacts between lower-level officials of all countries. The "corrective actions" the State Department recommended for saving money could not be accomplished by their suggested dates, Underwood argued, "unless arbitrary, unilateral action is desired," and he desired none of it.[43] Further, the commander warned ominously against clumsy attempts to reduce the size or scope of these tours "in order to minimize predictable unfavorable host country reaction to yet another case of what may strike them as paternalistic advice."[44] Underwood believed he knew best how the Informational Program affected its targets, and he deflected these attempts to curb his initiative by cannily citing traditional Latin American critiques of overbearing American behavior the State Department was sure to recognize.

In directing the study, assistant secretary of defense Nutter invoked the Nixon Doctrine's emphasis on "enhancing the capacity of nations receiving US military assistance to assume their own defense burdens" to justify close examination of the IP, presumably to find cost savings.[45] The US Southern Command commander deflected this critical examination by reinterpreting the Nixon Doctrine back to Washington—in essence claiming the authority to decide for himself the strategic balance between costs, measurable results, and intangible psychological benefits in the form of Military Internationalism that could be counted on to reap benefits sometime in the future.

Despite the supposed, unmourned death of modernization theory, many of the same older, deeper assumptions about the nature of American society and its power to bring about sweeping global transformations that had originally crystallized around modernization theory survived among those Americans who administered foreign military training programs.[46] Those who did often displayed a key paradox of the military brand of internationalism: its participants freely adjusted its ideological content the further one traveled from the national centers of foreign policy and toward the points of interaction between individuals of different nations. Americans such as General Underwood, or volunteer sponsors from around military facilities like the Greater Kansas City Chapter of People-to-People International, felt empowered to reinterpret the ideological content of programs that produced these interactions according to their own understanding of the national and humanitarian stakes involved. In this case, General Underwood perceived that a rash, Washington-directed plan to slash a popular IP perk—as a grand US tour for two hundred to three hundred senior South American military officers seemed to be—violated his own understanding of the goals of American military assistance policy in Latin America. Proximity to those South American officers informed his perspective on the exchange program, but his comfort objecting to the proposed cuts in the terms he did demonstrated that he exercised oversight from an ideological level that he felt empowered to define for himself.

The US Air Force's director of military assistance and sales, Colonel Atlee R. Ellis, also objected to several of the State Department's recommendations in ways that further reveal Military Internationalism's local, versus national, manifestations. Specifically, Ellis rejected the report's recommendation that the Department of Defense closely administer all IP programs and "provide the Military Departments with central policy guidance."[47] Program standards that ignored the diversity of foreign officers and the air force–run schools they attended would be awkward and, perhaps, humiliating, Ellis argued. It was more than a case of one size not fitting all. "The vitality of the program," Ellis responded, "stems from the initiative and the ingenuity of individuals in the field who have the task of making the program function." Neither these inspectors nor the Department of Defense seemed to understand the partnership between the official version of the IP and the communities that supported American military schools. "In many cases," Ellis wrote, "the financial obligations for picnics, luncheons, invitations to sporting events, visits to local industries, courts, newspapers, labor meetings, and other activities leading toward the achievement of IP objectives are borne by non-government, local and civic agencies providing the service for the trainees."[48]

Day-to-day operations of diverse IP activities at military training facilities around the country were too closely connected with informal community-sponsored activities to make it possible to separate the two for the purpose of standardization or, it would seem, regulation. Ellis's objection to standardizing the program from a central location corroborates that paradox of Military Internationalism. He only appeared to advocate the "wider use of private initiatives" the Peterson Committee recommended as a "New Approach to Foreign Aid."[49] Assumptions about the impact of life in American communities on foreign military visitors preceded the Peterson Committee, just as they had modernization theory. Official justifications and administration of international exchange programs grafted themselves onto community-based projects that already existed, in many cases. New approaches in foreign policy adopted these assumptions, but they did not create them. The State Department inspectors had recognized the same effect concerning the sponsor family program, which preceded and would likely outlive the official Informational Program. American officials came to understand that they were rarely in sole charge of the functions, funds, or ideology of military exchanges at the community level.

* * *

The first serious attempt to evaluate how the specific aspect of American military assistance programs that paired official training curricula with the extracurricular "lessons" in American culture and values revealed both continuity and adaptability. When American assistance planners in the early 1960s first articulated what they hoped to happen when military personnel from the Global South attended training in the United States, they focused on the intangible, psychological transformations bound to ensue in the wake of personal contact with American families. The international military education and training program not only survived the decline of "high modernization" in the late 1960s, it also kept alive many of the assumptions about the transformative power of that contact. American foreign policy de-emphasized direct military involvement in the target states, such as Ghana, as policy makers and scholars increasingly accepted that military regimes were bound to persist among American allies. As a result, faith in the transformative power of international military education and training and the Informational Program seemed as relevant as ever. Even when the Defense Department and State Department acknowledged that there was no way to measure the programs' effectiveness, they never seriously questioned their basic assumptions. Instead, they revealed an ongoing faith in the power of personal transnational contacts to undergird American foreign policy.

## Military Internationalism, Human Rights, and the Indestructible International Military Education and Training Program

Through the 1970s and again in 1979, Congress reassessed the purpose and execution of American military assistance programs, increasingly concerned with American military assistance for military regimes that violated their citizens' human rights, especially in southern Africa and Latin America. As this section explores, lawmakers also routinely questioned whether the United States had any evidence that its military assistance programs were worthwhile. They usually received similar answers: visiting military officers obtained an appreciation for the "American way of life" that made them less likely to abuse human rights, and American programs yielded long-term benefits far beyond the cost of their training. Specific variations on that theme over the years reveal different reasons why this was so, however. With savvy, the Defense Security Assistance Agency modified the rhetorical shell protecting international military education and training to deflect criticism connecting this specific assistance program and human rights violations abroad. As the next section will show, the response of American planners to the political situation in Ghana after 1979 demonstrates how flexible the rhetorical shell had become.

* * *

In the mid-1970s, the specific administrative and legislative definitions around international military training congealed, but the programs' ideological justification remained as fluid as ever. In 1976, Congress extracted the US training program from the larger Military Assistance Program and renamed it International Military Education and Training.[50] After 1975, the IMET program grew, from $29 million in 1975 to $56 million in 1985.[51] Between 1979 and 1982, IMET made up over 25 percent of the entire American military assistance budget, but as military assistance grew to $9 billion in 1985, IMET returned to below 7 percent of the total.[52] Before 1976, about half of the 430,000 students the United States claimed to have trained since 1950, including a majority of Southeast Asian and Latin American students, received training at home or in nearby countries, especially US schools in the Panama Canal Zone.[53] That training consisted of a full range of technical, administrative, and combat skills, and even after the Canal Zone school moved to the United States in 1984, exposure to "the American way of life" as prescribed by the Informational Program never really applied.[54] Most other IMET students—including all but 7 of 1,255 from West Africa before 1973—attended training programs in the United States.[55] In the 1970s, most foreign military students attended American courses in advanced professional education (especially staff colleges for officers and senior enlisted persons, such as the Command and General Staff College) and "resource management" (logistics, finance, administration, vehicle, equipment, and aircraft maintenance).[56] The State and Defense Departments administered IMET jointly and did not require a "specific military objective" as a condition for any state's participation.[57] While there was no requirement for military education and training to occur at military facilities, it usually did.

In the mid-1970s, defenders of education and training programs had to distinguish their program from other forms of American military assistance that had caused a string of humanitarian disasters, especially in Latin America. Reiterating their cost-effectiveness was one strategy. In March 1976, testifying before the House Committee on International Relations, Lieutenant General Howard M. Fish, director of the Defense Security Assistance Agency, stated, "Training of foreign military personnel is the most lasting value of military assistance and provides a simple, effective and relatively inexpensive contribution to the military strength of the free world."[58] The $32 million budget for international military training and education—about 10 percent of the total for military assistance—"multiplied manyfold" in results. The program augmented US foreign policy "by providing a significant opportunity for communicating with the military leadership of other countries, especially the emerging leaders and youthful elements of the developing nations," Fish argued. "The participants share the benefits of their experience with their countrymen when they return home to positions of responsibility and influence," he added. "Thus, while improving indigenous professional skills and competence, training at the same time serves the broadest US military and political interests and in all probability provides the greatest return on any portion of our military assistance program."[59] In a dispatch from Accra just four days earlier, US ambassador to Ghana Shirley Temple Black had repeated to Secretary of State Kissinger nearly the same points about the American exchange program with Ghana.[60] Behind any strategic or ideological justification for international military education and training, then, claims that it was inexpensive, cost-effective, and provided "a good return on investment" had survived since 1959 with little variation.

How to calculate that return remained elusive. In March 1976, undersecretary of state for security assistance Carlyle E. Maw repeated to the House Committee on International Relations the standard justifications for American training and education aid: "Our request for foreign military training amounts to only $30 million"; "The returns far exceed the costs"; and "Education of foreign military officers in American command, management and defense doctrine can contribute to personal international ties and understanding which will outlast any of the military equipment now being transferred to foreign governments."[61]

Congresswoman Helen S. Meyner (D-NJ) pressed back. "Even though this is relatively a small amount, $30 million when we deal as we do here at the Federal level in billions, why is this such good public relations?" she asked.[62] "Well, I think the understanding that foreign military people obtain of the United States and our way of life is a very worthwhile investment," Maw replied. "We bring promising officers to this country for broad training and experience in the United States. This has a lasting benefit." He went on, "My own feeling is that it is a tremendously valuable investment, something akin to the Fulbright scholarships where we bring people to the United States and give them exposure to our ways and methods and then they go back and take an important part in their own governments."

Meyner was incredulous. "You say they learn about our way of life. It seems to me they learn from us how to kill each other. I think they would be better off if we took them on a tour of Disney World, they would learn more about our way of life." "Well, that is perhaps a bit of a callous approach," Maw answered. "Most of these people are administrative. The facts of life are that in many countries of the world the military are the conservative forces helping maintain stability in their country." "Isn't it possible that we may be contributing by doing this to the decline of democracy in Latin America?" Meyner persisted. "Don't some of these people go back and overthrow their governments? Look what is happening in Argentina. Look at what did happen in Chile. This is what concerns me about this program. At least if they are going to have it, shouldn't they pay for it?" Maw deflected, "I don't know that the people trained in the United States are the ones that go back and try to overthrow governments. That would make an interesting study. I would think the training here would be one that would lend to stability rather than instability. I would prefer to have the military training in this country rather than in Eastern Europe, for example."[63]

Meyner's concern about whether American training was to blame for military officers who overthrew their governments after returning from the United States was familiar. Even defenders of the IMET program acknowledged that a rash of Latin American military rulers had received American training at the School of the Americas in the 1970s and 1980s, including General Omar Torrijos and General Manual Antonio Noriega of Panama, Major General Guillermo Rodriguez of Ecuador, Major General Juan Velasco Alvorado of Peru, Lieutenant General Roberto Viola and Lieutenant General Leopoldo Galtieiri of Argentina, General Hugo Banzer Suarez of Bolivia.[64] Fort Leavenworth's Command and General Staff College had also produced its share of graduates who later overthrew their government, just as Acheampong did in Ghana in 1972. In Kansas City, Ghanaian brigadier Ignatius Achel shared a sponsor family with Sudanese lieutenant colonel Jaafar Nimeiri, who seized power in a military coup in 1969.[65]

Nimeiri's ongoing relationship with his sponsor family throughout the 1970s reveals that, at least to the military travelers and the Americans they met, military rule may have been regrettable, but it was not always shameful. In 1970, Nimeiri personally invited Helen and Robert Black to Khartoum to help celebrate the one-year anniversary of his "revolution."[66] In 1972, while acknowledging that the new Ghanaian leader had also graduated from Fort Leavenworth's Command and General Staff College, the *Kansas City Times* recognized the Blacks with ambivalence.[67] It exposed, but refused to condemn, the many military leaders that had returned from Fort Leavenworth to assume positions of power in their governments. On a tour of the United States in 1977, Nimeiri arranged for Helen Black to join him on his presidential Boeing 727 jet for tours of San Francisco, New York, and Los Angeles. After a private tour of Disneyland, Black joined Nimeiri for dinner with Mayor Tom Bradley and actor Sydney Poitier at the home of African American music mogul Clarence Avant.[68] But as they had since the mid-1960s, while not specifically denying the downside, defenders of the

American international military education exchanges emphasized the potential benefits of US training. Until recently, few scholars argued that American military training made military men more likely to stage a coup.[69]

In 1976, critics grappled with the program's legacy. Under Secretary of State Maw parried Congresswoman Meyner with familiar ambiguity: nobody really knew the specific results of the program, but they agreed that it did good and provided a good return on investment. He also equated international military training with purely civilian educational exchanges, as if the Fulbright program and IMET served the same ideological purpose. In fact, defenders of international military education regularly fell back to this position, holding up cultural exchange and internationalism as self-evident results. Later that day, as he fed John E. Murphy of the Agency for International Development softball questions on the United States' concerns with foreign aid, Congressman John H. Buchanan Jr. (R-AL) illustrated a similar willful misunderstanding of American assistance programs that seemed to protect them from serious scrutiny: "Don't you think it is true," Buchanan asked Murphy, "that there is a great public misunderstanding of what development assistance is all about?"[70] "There is no question about it, Mr. Buchanan," Murphy replied. "You put your finger right on the nub of the problem—the lack of understanding on the part of the American people and the taxpayers of exactly what we are trying to do with the funds that we request for development assistance."[71]

This 1976 conversation echoed those of 1961, arriving at a similar conclusion: if the American people understood, Buchanan and Murphy agreed, then they would agree that this was money well spent because it matched their assumption that living among and learning from Americans provided visitors lessons in modern politics, economics, technology, and culture that, Americans assumed, visitors would carry home with them. This assumption, fulfilled at low cost, had changed little from when the Draper Committee had articulated it in 1959, and it survived the rise and decline of the modernization theorists and recurred at congressional hearings periodically throughout the 1970s and 1980s.

Congressional discussions about the past, present, and future of international military education in the late 1970s repeated similar questions about the potential negative consequences of American military training, its "return on investment," and how it fit into larger US foreign policy. In hearings before the Senate Committee on Foreign Relations in April, 1978, undersecretary of state for security assistance, science and technology Lucy Wilson Benson repeated, "This program supports foreign policy objectives by providing an effective and relatively inexpensive contribution to the military strength and leadership of certain countries."[72] She claimed that this professional training was "shifting in emphasis" to "broader fields of leadership training and resource management for senior military officers and junior and middle grade officers having leadership potential." Now, she said, training at "US military schools, e.g., Naval Command, Army Command and General Staff, and Air Command and Staff Colleges," focused on leadership and professionalism, rather than training mechanical and military skills. But this was not entirely accurate.[73]

Especially in Africa, as the assistant secretary of state for African affairs, Richard M. Moose Jr., testified, the line between "technical" and "professional" training was never clear. "Our proposed international military education and training programs this year will emphasize the technical training of African military officers in our military schools," he stated.[74] Yet "the bulk of these funds will be used to provide training in professional management rather than equipment-oriented training." Recipient states, in other words, had their own purposes for consuming American military assistance. Even the program's three official objectives (improving international relations, teaching human rights, and improving national self-reliance) encompassed a wide variety of American military schools. Given this vagueness, the first principle—that life among Americans could transmit the most important lessons of all—still reigned. As the 1970s wore on, some in Congress and presidents Ford and Carter increasingly emphasized such ideological outcomes for American foreign policy.

In his 1975 article "The United States in Opposition," ambassador to India and future senator Daniel Patrick Moynihan defended American liberal internationalism and economic liberty against increasingly threatened political and civil liberties abroad.[75] He announced the United States' new commitment to human rights to the world and likely earned President Ford's nomination to United Nations ambassador in appreciation.[76] Neither Moynihan nor Ford was first to champion what later became the most distinguishable aspect of Carter's foreign policy.[77] Even Henry Kissinger accepted that values, in one form or another, had always been a central concern of American foreign policy, and that most Americans considered the amoral pursuit of power very un-American.[78]

The State and Defense Departments anticipated increased congressional scrutiny of military assistance programs' spotty records of supporting dictators in the mid-1970s, and they responded by touting their cost-effectiveness. Without irony, they also reformed the mandate for international military education and training to stake a claim as a primary method of instructing an American version of "internationally recognized human rights."[79] Doing so came naturally, since the original assumptions about what made military education and training in the United States effective already affirmed the enduring value of human-to-human contact and Americans' particular excellence at it. At least on paper, encouraging human rights became equal to the IMET program's other main purposes of encouraging effective and mutually beneficial relations between the United States and target states and improving those states' ability to utilize their own resources (which might include defense articles the United States had sold them). In 1978, Congress amended the Foreign Assistance Act of 1961 to forbid all military assistance—including IMET grants—to states that engaged in "a consistent pattern of gross violations of internationally recognized human rights" and made "increas[ing] awareness of nationals of foreign countries participating in such activities" one of the IMET program's three main objectives.[80]

Even the new emphasis on "human rights issues as part of this leadership training" that Under Secretary Benson cited blurred lines between theory and practice in international assistance. The Department of Defense took until 1985 to update its directive on the Informational Program to reflect the added emphasis on "internationally recognized human rights," but the department included this emphasis in its rehearsed and recycled repertoire of justifications for IMET routinely after 1978.[81]

As Congress debated the International Security Assistance Act of 1978 that eventually prevented American military assistance to regimes that violated human rights, President Carter directed the Pentagon to review the policies and procedures governing US training of foreign military students. This study intended to ensure the program followed a coherent rationale, that it reflected changing American foreign policy objectives, and that it allocated funds according to larger military and political priorities.[82] The General Research Corporation published its thorough, if uninspiring, report in March 1979. Like the 1971 review, this one included the standard articles of faith about the psychological impact of training in the United States on foreign military visitors and on the program's benefits for American influence abroad, and it recommended vague policy changes to make the program more responsive to various inputs.

From a different angle, however, this report reiterated in stunning detail how American political and diplomatic assumptions behind international military education interacted with individual military schools and training facilities and with local communities to imagine a set of psychological outcomes in target audiences that mirrored American foreign policy objectives. All of these variables interacted in specific locations (schools, bases, communities, etc.), but the logic of the overall system envisioned a universal process of transformation to create the desired outcome because the definitions surrounding American military education and training programs were so pliable. Finally, the report revealed the primacy that American policy makers gave to national formation as the desired outcome of its international military training and education programs. But in the processes the report describes, nationalism was only an immature form of internationalism activated through military travel. National actors provided only the initial conditions for the transformative power of the educational experience in the United States. The desired policy-level outcome of nationalism failed to recognize that American and foreign individuals themselves determined the actual psychological impact of the international exchange.

The Defense Security Assistance Agency, which commissioned the 1979 study of international military education, considered its justification self-evident, and the report revealed familiar critiques. First, the study recognized (as this chapter has), "the rationale for IMET appears not to have been uniform, vacillating in time from one concept to another."[83] This report attempted to devise a standard rationale, as well as a standard method for prioritizing allocations, but even in its pragmatic introductory summary, it repeated its broader faith in the United States' ability to transform military visitors psychologically. "All IMET serves some specific and legitimate military purpose within the armed forces of the trainee's country," the report claimed, and "IMET is not a cultural exchange program, although the exchange in cultural values inherent in bringing foreign and US trainees together may be an additional benefit."[84] This exact phrase occurs repeatedly in congressional documents over the next decade, usually with the addition of "an additional cost-free benefit."[85] Further, since most of the training occurred in the United States, English became trainees' second language, "opening the door to Western literature well beyond military subjects."[86] Even when attempting to be pragmatic, these observers waxed philosophical.

Defenders of professional military education acknowledged that even the term "professional" defied quantification. US policy makers had been telling one another that their international military education and training programs emphasized "professional" courses, such as staff colleges, over mere "technical" training, but by 1979, there was still no clear definition of what made some military training "professional" and some not.[87] For Ambassador Shirley Temple Black, for example, "professional" courses might include ranger, airborne, flight safety, and helicopter pilot training in addition to staff college, but staff college slots were the most sought-after.[88] The Congressional Research Service (CRS) acknowledged, "A precise definition of the term 'professional military training' is not available at the US Government interagency level, and there are varying views as to exactly what is meant."[89] To some officials, the term implied "postgraduate education of the type to be had in civilian institutions"; to others, the term included "certain courses of instruction for officers, particularly higher level courses, such as command and staff or war college." Clearly staff colleges deserved the distinction, but there were also numerous other courses in the United States that were "difficult to so differentiate, and their arbitrary classification appears to serve no real useful purpose."[90] It might be more meaningful to simply define professional military education as "that training designed to provide or enhance recipients' leadership"—not a difficult standard to argue.[91] At best, only a very broad definition for "professional military education" would be feasible, the report suggested. "Better still," it recommended, "for purposes of IMET, *no definition at all*."[92] Officers from "the less sophisticated countries have little need for what might be termed war college–level professional military training," the report concluded, since their own training bases did not "produce personnel sufficiently trained to absorb higher level training."[93]

These analysts' struggled to come to terms with what made "professional training" count as "professional," or even useful, for foreign visitors, and they revealed part of the paradox of Military Internationalism. Not only did they decide it was impossible to clearly define "professional training" in a way that included both American and foreign assumptions, they decided it was best to *avoid* defining it at all. What was professional about any American training course would speak for itself, they assumed, because it was the environment of the course itself that was transformative, not the content of the training. Rostow had said in 1961, in essence, that international military training was too important to be left to the generals. In 1979, the CRS recommended against letting the definition of "professional" interfere with the intent of professional military education for foreign visitors—especially ones from "the less sophisticated countries." In this paradox of Military Internationalism, definitions were not only changeable, but their very presence hindered the possibility for cooperation. From the point of view of the recipient countries, a rigid set of American definitions limiting which schools were available for grant tuition may also have restricted their ability or willingness to seek this grant aid.

Indeed, the purpose of international military education and training, as the United States saw it, was rarely to produce any specific military objectives at all. By legislation, the program's three purposes were to encourage mutual relations between the United States and client states, to provide military training to help them utilize their own military equipment effectively in order to become self-sufficient, and "to increase the awareness of foreign nationals of basic issues involving internationally recognized human rights."[94] This report's conclusions about the inability of "the less sophisticated countries" to absorb the "higher level training" in some American courses and the legislative mandate to "increase awareness" regarding human rights also reveals the paradoxical relationship between the ideal versus the actual militaries in question and the ideal versus the actual states involved. As Ghana's ongoing participation in American military education and training programs throughout its period of military rule shows, Americans' commitment to international military education and training tolerated remarkable diversity in client states' domestic political and economic development or military sophistication. While the language in this particular report sounds condescending, the larger paradox of Military Internationalism was that it accepted considerable discontinuity between ideal political forms and civil-military relationships, and those forms and relationships actually in place in client states.

American assistance planners' version of Military Internationalism did not hold client states to particularly high standards of governance because they considered military education in the United States itself the method through which to transform those states. To some extent, the American attitude was one of tolerating friendly dictatorships abroad, even those that scholars legitimately considered harmful to democracy and justice in the Global South.[95] But from another perspective, the Americans responsible for executing this particular aspect of US foreign policy continued to believe, as they had for two decades, that their program would be the mechanism through which democracy or justice might come to those states, via the military.

Conflict over the centrality of the nation-state underlay the assumption that international military education and training yielded impressive, if vague, benefits. In part reflecting 1970s-era tolerance of bureaucratic authoritarianism, the 1979 report revealed a bipolar attitude toward national cooperation for its own sake versus for the sake of power. "In the simplest of terms," its authors asserted, "the world is a collection of nation states. . . . The nation state remains the primary actor in the world, despite the greater ease of international travel and communication, despite greater interdependence between nations, and despite the advocacy in some quarters for a more interdependent world."[96] The United States was a nation-state "seeking to survive in a world system animated by considerations of power," so it must combine its power with other nations "in order to ensure its own survival."[97] Foreign military personnel attended US military training institutions to "acquire information and insight . . . on which to base a decision as to the practicality of cooperation with the US on military matters."[98] Therefore, "When a foreign country agrees to send its military personnel to be trained by the US, it casts a vote of confidence in the US and its military institutions. . . . When the US offers training to foreign military personnel on a grant basis, it demonstrates a continuing real and active interest on the part of the US in the national security of the foreign country."[99] Further, since IMET aid was a grant program, the United States maintained some authority over which individuals could receive the training.[100] According to these justifications, considerations of state power and national interest sustained these programs, so their effectiveness rested on how well they achieved these goals.

Questions of modernity and the institutional connection between poverty and development became even more pressing in the 1970s, but relieving poverty or bringing modernity to the Global South are notably absent from this 1979 explanation of the intent of American military assistance.[101] The way IMET depended on a visiting individual's personal conversion to faith in American political, economic, and social institutions provided an alternate mechanism to conform to new directions in foreign policy in the 1970s that tolerated some bureaucratic-authoritarian regimes while de-emphasizing some aspects of state power—such as the nagging persistence of military governments among American allies. Even as the discourse of aid shifted to de-emphasize state-based projects and embrace relieving individual poverty as a condition of assistance, IMET continued to declare that its original genius was to transform individuals, who in turn modernized their societies' proto-national military institutions.

Policy makers in 1979 still could not comprehend how to measure the success of international military education and training according to nation-based criteria. They accepted as evidence that "large numbers of students that friendly foreign governments have been willing to commit to the armed forces of the US for training and the number of countries willing to adopt, in large measure, the tactics, techniques, and weaponry of the US are indicative of the success of IMET and prior grant programs."[102] This is a weak endorsement, however, considering that few other states offered comparable assistance, and the United States specifically recruited client states away from those that did—usually from the Soviet bloc.[103] In 1976, Secretary of State Kissinger specifically asked all foreign embassies whether reduced US IMET aid would encourage those countries to look to China or the Soviets—and the US ambassador from Ghana replied that it probably would.[104] The large number of graduates of American military training courses who had advanced to positions of prominence in their home militaries or home states offered another way to demonstrate the success of the programs.[105] On the other hand, many illegitimate world leaders had also graduated from American military schools—although this report mentioned few by name. The specific process for budgeting IMET training changed in the 1970s, and IMET programs received their own specific appropriation separate from the larger Military Assistance Program. The programs' lack of concrete results made them increasingly vulnerable to arbitrary cuts by Congress, "despite Executive Branch arguments that the training of foreign military personnel is one of the most effective means of enhancing the security and influencing the policies of recipient countries."[106] Terms of nation-states and power still failed to illuminate these programs' tangible results.

Where this report attempted to describe how military education and training programs advanced American geopolitical interests, it extolled the intangible benefits of international exchange to make its case. "It is in the training environment that military-to-military relationships of enduring value to the US are made," the report claimed. More than that, it said, "when US and foreign military personnel sit down to plan and undergo training together, the mutual insight and rapport that result are *of a different and higher order* than relationships based on most other types of contact."[107] It argued that the American people continued to support international military education and training programs despite their unstable rationale, their vulnerability to budget pressure and "political considerations," and their tendency to "vacillate from one concept to another," because they directly participated in these relationships.[108]

The programs evaded measurement because they were pluripotent. To Ernest Lefever of the Brookings Institution, whose earlier work contributed to the Defense Security Assistance Agency's 1979 report, the true benefits of foreign education and training programs took decades to mature. In the short run, some American-trained officers initiated coups in their home states, but others fought against them; some advocated the nationalization of US corporations in their countries, others resisted it; in civil wars both rebels and loyalists had trained in the United States.[109] The real benefit of the programs was more like an institutional subconscious, where "even if a small number of these attempt to emulate the Democratic values and practices they heard about and saw in the United States, the net, long-term effect, however modest, would be in the direction of a more representative and responsive government than most third world countries now enjoy."[110] It might take decades for the fruits of these exchange programs to mature, these analysts reckoned, but that slow, psychological change would be more meaningful than any short-term policy alignment between the recipient nation and the United States—especially when those policy alignments rarely occurred as the United States hoped.

The General Research Corporation's American perspective did not acknowledge the power differentials between American training staff and foreign military students, and it assumed that the individuals involved in these courses viewed themselves primarily as *national* beings when participating in them.[111] In practice, differences in power and nationality required constant negotiation. On post, among men and women in uniform, power differentials could hardly be hidden, and only American faculty and senior officers had the authority to downplay them between fellow students or between students and instructors. In social situations, where community members struggled to host welcoming and inclusive events, specific "national" labels adhered with less uniformity. But the policy discussions surrounding these programs justified them as pragmatic national policies while at the same time touting the enduring personal bonds they created.

In such a way, these programs reveal another paradox of Military Internationalism: its uneven dependence on state power. States facilitated international exchanges and provided official justifications to sustain them as government programs. After arranging these contacts, however, pragmatic rationalizations rooted in foreign policy objectives and state-building gave way to a deeper faith that visiting foreign military personnel and American families would interact in ways that represented a more natural internationalism that was not solely about national power.

By defending international military education and training as a central site for transmitting American values and "ways of life," Defense and State Department planners appealed to international norms for practical as well as ideological reasons. Continuously after 1979, staff in both departments claimed to "weigh human rights practices in each of the proposed recipient countries" and argued that international military education and training best formed apolitical and professional graduates who, in turn, supported "democratic processes" and "the basic rights of individuals."[112] Emphasizing that they more than complied with the International Security Assistance Act of 1978's prohibition against IMET grants to states that grossly violated human rights, they preserved their program from excessive scrutiny and offered it as a proactive American endeavor to spread human rights awareness. In a kind of reverse "boomerang effect," policy makers employed the internationally recognized rhetoric of human rights to defend the program they considered essential for forming transnational networks of military officers who, they argued, could bring about the reforms human rights discourse prescribed.[113]

One might question their sincerity, or at least recognize the absence of evidence for their point, but by regularly touting the way life in American communities actually improved foreign military personnel, policy makers appealed to deep-rooted assumptions. They were the same assumptions that motivated reformers of the old Mutual Security Program of the 1950s to replace the old justification of containing communism with a new emphasis on delivering political, economic, and, most of all, psychological modernity to target societies in the Global South.

* * *

US and foreign governments employed a version of Military Internationalism that professed great faith in the power of international travel and mutual education and training, and they operated those programs according to corporate calculations of national political, geopolitical, and economic interests. But in order for those contacts to yield "national" results—results based on the participants becoming more national beings by serving in their nation's armed forces—very *non-national*, human interactions had to occur first. In this way, state-based explanations of their programs still occasionally referred to the transnational benefits of interaction. As a state-supported project, these programs adhered to a version of internationalism that privileged the "national" results of the exchange. The 1979 evaluation of IMET repeated the potential for national development and increased "access" for American diplomacy in the Global South. But a different, more cultural understanding of the consequences of exchange lay under the administration of international military education and training programs, which observers obliquely acknowledged surprisingly often.

## The Elastic Rhetoric of International Military Education and Training: The Ghanaian Case

Throughout the 1980s, the language the Department of Defense used to defend international military education and training programs routinely stressed the pragmatic, national outcomes the programs delivered while acknowledging that they resulted from non-national, cultural, and human interactions. Official US justification for IMET grants to Ghana during this time demonstrate the nexus between pragmatic national interests and a near-mystical belief in human-centered psychological transformations particularly well. Ghana's case also demonstrates how the vocabulary around American military assistance programs adroitly shifted its shape without greatly disturbing the rhetorical or institutional inertia that supported them. In annual congressional presentations on security assistance programs from the 1980s, descriptions of events in Ghana demonstrated the flexible rhetoric defending the transformative power not just of American military assistance, but of international military education and training specifically. When the Ghanaian armed forces as an *institution* failed to behave in ways American planners predicted it would when under stress—as it was in the years surrounding the 1979 Rawlings revolution—the Defense and State Departments fell back to their previous justifications for IMET's impact on *individuals.*

* * *

Through the 1970s and 1980s, the United States' only security assistance to Ghana was through IMET. The program funded an average of thirty Ghanaian personnel to attend training or education programs in the United States and cost the United States about $240,000 per year.[114] Most, like Ignatius Acheampong, Henry Kwame Anyidoho, and Ignatius Achel, attended American staff colleges; the rest attended programs for "military management and technical areas such as finance, administration, and engineering."[115] Annual congressional presentations after 1979 did not state which specific schools each Ghanaian attended in the United States; instead, they all generalized around the theme of "improv[ing] Ghana's own indigenous training capability and upgrad[ing] the educational and training standards of armed forces personnel."[116] Compared to approximately $15 million per year in other types of US economic assistance to Ghana between 1978 and 1984, (including P. L. 480 "Food for Peace" food aid, Agency for International Development assistance, and the Peace Corps), IMET was a small program, but the Defense and State Departments considered it the United States' only avenue of potential influence over Ghana's politically significant military.[117]

The Department of Defense and State Department repeatedly claimed that the small IMET program in Ghana guided the Ghanaian government's rapidly evolving relationship with the United States after 1979. As expected, the Defense Security Assistance Agency's congressional presentation for fiscal year 1980, written early in 1979, proclaimed that IMET "serves as an important element in our friendly relations with the Ghanaian government by providing professional training to key military leaders, current and potential," while boasting, "the US experience has also created understanding and respect for American values and institutions among Ghanaian military leaders."[118] The DSAA also suggested that US military training stabilized Ghana during uncertain political times. "In June 1978," it recalled, without recognizing that Acheampong himself was an IMET graduate, "military officers led a bloodless coup that replaced the government of General Acheampong. The new government has released political detainees and undertaken difficult and long overdue economic reforms. It has also upheld the previous regime's commitment to step down in favor of a democratically elected civilian government."[119] The report projected that even after political power transferred to civilians, "Ghana's military can be expected to play an important role in the country's economic and political development."[120] In early 1979, Ghana was on the verge of great political change. The DSAA intended that American-trained military officers would provide a steady, human rights–respecting influence and reliable partnership with the United States.

The Defense Security Assistance Agency's 1980 congressional presentation mistakenly projected an even larger role for the armed forces in stabilizing Ghanaian politics.[121] When this report arrived, Flight Lieutenant Jerry J. Rawlings's Armed Forces Revolutionary Council (AFRC), which toppled the Supreme Military Council in June 1979, had temporarily relinquished power to a civilian government under President Hilla Limann.[122] Rawlings later overthrew the Third Republic on December 31, 1981, but in the interim, the DSAA asserted that Limann's government would depend on the "cooperation and support" of the Ghana Armed Forces for its survival.[123] The reasons for the 1979 coup by "rank and file military," the DSAA reported, were "severe inequities between officers and enlisted men, corruption in the military and the country's depressed economy." Although the report noted that Rawlings's AFRC had executed eight top leaders, "including three former heads of state," it did not mention Acheampong among them.[124] It also failed to evaluate whether a decade of American professional training had mitigated the inequities or corruption. Meanwhile, the United States' "modest security assistance program" would remain "an important element in our friendly relations with the Ghanaian Government by providing professional training to key military leaders, current and potential."[125] In 1980, and again in the fiscal year 1982 report (prepared in 1981, before Rawlings's second coup), the Defense Security Assistance Agency boasted, "The professional training offered key members of the Ghanaian military in the US has given them an understanding of and respect for American values and institutions."[126] Further, the report suggested, "this reinforces their Western orientation and contributes to strengthening the friendly, cooperative relationship between Ghana and the US." These are standard justifications, based on the oft-repeated narrative about the long-term psychological and diplomatic benefits of American training. At this point, despite the illegitimate rise and violent fall of one IMET-trained head of state in Ghana, the Ghanaian experience did not demand that the United States reconsider international military education and training as a whole.

When preparing the fiscal year 1983 congressional presentation in 1982, after Rawlings returned to power, the Defense Security Assistance Agency acknowledged Ghana's new military government, but the agency either failed or refused to acknowledge its revolutionary character.[127] In January 1982, Rawlings declared "revolution"—he later called it a "holy war"—and accused Ghana's military of "years of corruption in the highest ranks which has also seeped down the command structure."[128] He declared: "We have now a People's Army, a People's Navy, a People's Air Force. . . . We have seen enough of a traditional form of leadership which abandons the people once it is in power."[129] "The Revolution has no room for elitist, self-centered and materialistic officers," the new army commander Brigadier Arnold Quainoo declared to Ghana Military Academy's 1982 class.[130] Meanwhile, Rawlings restored diplomatic relations with Libya, and broadcast a new willingness to cooperate with the Soviet Union, Iran, and Eastern Europe.[131]

Later scholars considered Rawlings's ascent to power a symptom of the military's inability to convert political dominance into institutional stability, while the charismatic leader merged the military's strong corporate identity with the radical aspirations of its lower ranks' (and of the Ghanaian poor).[132] Paradoxically, decades of international military training stunted the Ghana Armed Forces' ability to renew itself organically and contributed to Rawlings's grievances against the military hierarchy. Ghana's "complete dependence on international circuits of military production and training" constricted junior ranks' advancement.[133] American policy makers at the time interpreted the coup not as the failure of American attempts to reshape the institutional or corporate character of Ghana's military in the past decade but as evidence of an even greater need for "training to strengthen professionalism and discipline."[134] But this required a shift in the rhetoric supporting American military assistance away from its prior gradualist emphasis on influencing Ghanaian military development and toward restoring a lost, mythical professionalism.

The Department of Defense assumed that its IMET program had such power. Since 1979, the Defense Security Assistance Agency's annual reports had repeated familiar refrains: Education and training in the United States prepared foreign military leaders for "operating, maintaining and managing a military establishment," while reinforcing "a political and professional military element in support of democratic processes."[135] Training under IMET was "professional and nonpolitical, reflecting the US tradition of civilian supremacy and the instrumental rather than policy role of the military."[136] Defense policy makers still boasted, "IMET is not a cultural exchange program, but the exchange of cultural values inherent in bringing foreign and US military trainees together is an additional and essentially cost free benefit."[137] Meanwhile, the assistant secretary of state for African affairs affirmed, "Without exception, each of our ambassadors reports that IMET is one of the most valuable programs we have to offer."[138]

Recognizing the Rawlings government's "anti-US sentiments," in 1983, the Defense Security Assistance Agency declared IMET to be even more important and "one of the few remaining vehicles for fostering US interests in Ghana."[139] Rawlings and the junior officers and noncommissioned officers who conducted the coup "exacerbated problems of military discipline, and increased the need for professional training." In 1984, noting that US-Ghanaian relations "improved perceptibly starting in the summer of 1983," the agency reaffirmed Ghanaians' critical need for American military training, since "the trainees are exposed to American values and democratic, human rights oriented institutions, including the traditional role of the US military within our society."[140] The United States recognized that "although political developments in Ghana during the past few years have resulted in an erosion of the influence within the military of Western oriented personnel, there is still a significant amount of respect for Western techniques and systems." Finally, "Maintenance of the [IMET] program is one of the principal means of professional contact with the Ghanaian military at a time when military personnel of countries hostile to the US, i.e. Libya, are increasingly present in Ghana."

By 1985, after Ghana's economy had failed to recover and Rawlings had invited the World Bank to intervene and Ghana had adopted a US-supported structural adjustment program, the Defense Security Assistance Agency was optimistic about US assistance again. "The unquestionable professional benefits of the security assistance program are considerably enhanced by the opportunity to demonstrate to key personnel within the Ghanaian defense establishment the traditional role of the US military within our society," they reported.[141] Similarly, in 1986, as Rawlings's military government continued into its fourth year, the Defense and State Departments were sure their education and training program could "encourage nonalignment and counter Soviet bloc influence, and stimulate the development of the new democratic system."[142]

Political uncertainty following Rawlings's first coup in 1979 gravely challenged the traditional narrative justifying American military education and training grants to Ghana as a source of long-term stability. In response, US policy makers relabeled the program's purpose into an essential toehold for American interests that simultaneously preserved and retransmitted an ideal of civil-military relations—despite little historical evidence of success.

The program continued to sponsor roughly the same number of students for the next decade, but the rhetoric the United States employed to justify its programming decisions reveals a combination of underlying attitudes about international military education and Military Internationalism.

First, as an individual identity and not a schedule for corporate or national development, Military Internationalism preceded specific program descriptions and influenced how observers evaluated programs' results. In 1970, 1979, and again in 1990, congressional critiques of IMET chided its inability to "measure" results, but American policy makers never questioned whether sponsoring just twenty to thirty Ghanaian military officers' training in the United States each year would eventually yield society-wide changes. Instead, when a group of junior officers who had not yet participated in American professional military education seized the government, American policy makers reformulated their position from emphasizing IMET's conservative tendency to projecting that the new government required a transformation that Ghana was unable to provide for itself. In annual reevaluations judging and justifying IMET in Ghana after 1979, the Defense Security Assistance Agency and the State Department did not consider Rawlings's revolution evidence of the failure of a decade and a half of American professional military influence. As they did after previous coups, academic observers quickly recognized that Ghana's revolution had fundamentally Ghanaian origins.[143] Instead, military assistance planners proposed that international military education and training was the last best hope for Ghanaian governance and US-Ghanaian relations. Similar to the experience of failed modernization projects in the early 1960s, results from Ghana never seriously challenged the faith in the power of Military Internationalism inherent in international military education and training; they only prompted a more expansive expectation for what Military Internationalism might accomplish.[144] IMET's justification and application evolved in light of new conditions, but the underlying international identity remained unquestioned—at least by military assistance planners.

## Momentum or Inertia in the 1980s

The United States' International Military Education and Training Program changed remarkably little during the 1980s, even as fierce battles raged in US government and academia over the proper course for American foreign policy with respect to Asia, Africa, and Latin America. As scholars reassessed the persistence of military rule abroad, Congress continued to challenge American military assistance programs for nondemocratic and abusive regimes abroad.[145] Spokespersons for the IMET program from both the Department of Defense and the State Department deftly defended the program using language that embraced the long-held faith in the power of American communities to improve visitors from the Global South while offering long-term strategic rewards. Best of all, by weaving congressionally mandated concern for human rights into the curriculum for foreign visitors, client states' political, economic, and cultural transformations were bound to result as a "cost-free" benefit for US foreign policy.

* * *

The simultaneous growth of American interventionism and exceptionalism in US foreign policy during the Reagan years required security assistance programs to serve several purposes at the same time. Military and economic policy under Reagan combined a renewed dedication to intervention worldwide with the ongoing commitment to deregulation and rejection of the social corporatism inherited from the Nixon era.[146] Early in his first term, Reagan and his advisors pushed against weakening Soviet influence everyplace they could detect it through massive rearmament, applying economic and diplomatic pressure, and sending military assistance to counterrevolutionary movements and pro-Western regimes across the Global South.[147] American security assistance expanded sevenfold between 1980 and 1985, economic assistance abroad doubled, and the total American security assistance budget topped $9.7 billion.[148] IMET doubled in dollars to fifty-six million, but its portion in relation to other types of military assistance shrunk from 25 percent to just 7 percent and less than 1 percent of all American aid.[149]

US hegemony over the institutions of global finance also made economic aid contingent on accepting American neoliberal economic ideology. Especially during the global recession of 1981–1982, revolutionary regimes across Asia and Africa, including China and Rawlings's Ghana, increasingly tabled socialism in exchange for International Monetary Fund loans.[150] The Defense Security Assistance Agency marked these changes in Ghanaian domestic economic policy in their annual summary of American aims for IMET in Ghana, while asserting that the United States' ongoing program to train influential Ghanaian military officers moderated extreme fluctuations in Ghanaian domestic politics.[151]

Ironically, waging economic cold war in the Global South with the carrot-stick combination of military aid and economic assistance that demanded "conditionality" and "adjustment" to market forces exacerbated the United States' own domestic economic problems. By the mid-1980s, unprecedented annual budget deficits defied traditional American domestic political arrangements.[152] American liberals and conservatives alike attacked the massive foreign aid infrastructure, hotly contested US intervention in Latin America, and questioned military sales programs that, they feared, caused arms races and instability at the expense of American economic stability.[153] As they had since the 1950s, congressional debates over the size and purpose of the United States' security assistance programs revealed where individuals' regional and domestic concerns merged with discussions on international policy.[154]

Especially in the Democrat-controlled House, where representatives scoured the security assistance budget in search of areas to cut, the ideologies behind each military assistance program occasionally revealed themselves in high relief. In a typical example from 1985, David Obey (D-WI), chair of the Foreign Operations Subcommittee of the Committee on Appropriations squared off against retired lieutenant general Ernest Graves, former director of the Defense Security Assistance Agency, currently a senior fellow in International Security Studies at Georgetown University, and Frank Kramer, former deputy assistant secretary of defense for international security affairs, when looking to cut $1 billion from President Reagan's $9 billion foreign aid request.

"If we are going to be looking at a 10 percent cut, we ought to start with the big babies first," Obey opened. "What do we really get out of a $50, $75,000 IMET program in any of these places? Can I really come with a straight face, tell my constituents or tell members of the House on the floor, if I am asked, these dollars are essential this year—given the budget crunch—essential to the national interests of the United States?"[155] "Well, the goal in the smaller countries . . ." Graves attempted to answer. Obey cut him off: "I recognize what the goal is, and I am not questioning the fact that these programs might be nice to have or useful." The question was, he demanded, "if we are dealing with a budget crunch, are they . . . essential to the national interests of the United States?"

Graves and Kramer took turns responding. "I think the position of the administration in including these is that they see improving our security ties with those countries as fitting in with the overall effort to improve security around the world. And I think they are probably taking the view that the $50,000 is just as well spent as the $50,000 in some of the larger programs," Graves answered. "A critical point to remember about IMET," Kramer added, "is that it is a long-term program. It is going to have no immediate consequences that are essentially in the same way as putting a weapon in the hands of a country that is being attacked. So when you cut it, you'll not have immediate consequences." Congress may find itself with a "short-term" deficit problem, Kramer warned, and may find cutting IMET a tempting short-term solution, so "I think the answer to your question is no, it is not critical," he admitted. "But if you take a longer view, there are some countries, I think, that you can say would never be essential or are very unlikely to be, but the whole purpose of the IMET program is to create a long-term and enduring relationship."[156]

Defending the program despite the difficulty of measuring its effectiveness was only one challenge. "Some people do not like the fact that we have trained people who have ended up being dictators of various countries," Kramer admitted, "but the fact is we have incredible access because of that program, and probably dollar for dollar it is the most important program we run."[157] So, he summed up, "whether you have to have IMET in Guinea, I doubt it over a given year. I think you have to look at it over a given number of years."

This conventional response explicitly reiterated many of the rhetorical and ideological shields that protected international military education and training from serious scrutiny, and it implied the rest. IMET was a long-term program; critics must avoid the temptation to fixate on the lack of measurable results. Even for countries where the United States had no national interest, the program kept open channels of "incredible access" and permitted the United States to maintain more than just a token diplomatic presence nearly everywhere. More than that, the program built "relationships," both institutional and individual. At the time of this hearing, more than one hundred nations participated in the US IMET program in some way.[158] Even in Guinea—which in this testimony was shorthand for "the least important places of the Third World"—the United States could extend its psychological influence. Admittedly, many graduates of the program committed improper deeds—as this and other hearings throughout the 1980s confessed—but the program was always cost-effective.[159] Especially in a security assistance context dominated by massive grants to a few strategically significant recipients, the justifications inherent in IMET's modest involvement worldwide made it stand out as a form of military assistance that could please an exceptionally wide range of political constituencies.[160] In eras when budgets grew tight, the program's low overall cost and mythical cost-effectiveness actually preserved it from reductions.[161] These defenses had justified the program's expansion in the early 1960s, had survived the decline of high modernization and bureaucratic decentralization in the 1970s, and persisted through the 1980s.

Congressional attempts at the end of the 1980s to come to terms with IMET's cost-effectiveness and lack of measurable results revealed how persistent and consistent the ideals remained. In September 1989, Senator Patrick Leahy (D-VT) again acknowledged decades-old critiques of the program: "There has never been an objective evaluation of whether IMET training actually changes attitudes about respect for human rights and civilian control," he told the Committee on Appropriations.[162] "The Committee is told that IMET is a superb program with high payoff to the United States in foreign policy and national security terms," he said.[163] "If this is so self-evident, it should be easy for the appropriate agencies in the executive branch to prepare a full analysis of the program demonstrating the achievements of the IMET program."[164] In June 1990, as Leahy announced the publication of the General Accountability Office report on IMET that he had commissioned, he declared, "IMET is a program that runs on momentum or inertia. Until this GAO study I requested, I do not think that anybody had looked at IMET since the early 1970s."[165] He was not entirely correct—the Defense Security Assistance Agency published a complete review of the program in 1979, as this chapter describes. But Leahy's conclusion was valid: despite vast changes in global politics, the Cold War, and the objectives of American foreign policy, IMET's justifications had not significantly changed since the early 1970s.

The General Accounting Office raised familiar objections to the international military training program's administration in 1979, but the ideology behind the program remained firmly entrenched. Neither the Defense nor the State Department had established a method of evaluating the success of international military education and training, they failed to maintain contact with IMET graduates or track their progress through their home militaries, and local training officials frequently changed their courses' content and deviated from approved course materials.[166] On the other hand, the GAO acknowledged, international military education and training programs seemed to succeed at "expos[ing] participants to US values, citizens, institutions, and commitment to human rights."[167] Despite the critiques by Leahy and the GAO, IMET's ideological foundation that international military training had the potential to transform visiting military officers psychologically—at a low cost—remained intact.

The Defense Security Assistance Agency published its 1989 responses to congressional inquiries in an academic article in 1990. Spiro C. Manolas and Louis J. Samelson wrote, "the Administration takes the view . . . that IMET *is* an effective, low-cost component of the $8 billion global US security assistance effort."[168] At the then-current level of $47 million per year, they argued, the program provided "US access to and influence[d] foreign governments far out of proportion to its modest cost." On the other hand, they could offer "no rigorous proof either supporting or contradicting" whether American education and training efforts "actually can change attitudes in foreign counties toward such core US concerns as civilian control and human rights."[169] Repeating the remarkably resilient phrases about the programs' benefits and compiling anecdotal evidence and testimonials from graduates of US-sponsored training, the 1990 assessment demonstrated that faith in the transformative power of Military Internationalism still underlay American assumptions about its foreign military education and training programs.

The 1990 assessment still identified national outcomes as a primary goal of American assistance, but it did so in disembodied ways. "Nation-building is not an objective of the IMET program per se," it reported, "nevertheless, it is an important byproduct."[170] Emphasizing "technology, management, and technical skills," these programs "had a positive effect on the infrastructure of IMET recipient countries. The effect has been to stimulate 'nation building' which, in turn, has encouraged economic development." This assertion fails to explain how a human-centered training experience could influence the infrastructure of a country. It relies on an implicit understanding that the foreign officer's experience in a US military school or training facility equated to an institutional intervention between American society and his own—and it could just as well have been written in 1960. This is not simply wishful thinking, the case of one executive branch department attempting to justify its budget. This assessment revealed two important aspects of this small, ongoing project of US military assistance: First, the expansiveness of the American faith in the technical and practical superiority of American training that permitted it to merge so successfully with modernization theory continued to permeate IMET for decades. Second, in emphasizing once again the national outcomes of the individuals' training experiences, this assertion illustrates the paradox that state actors failed to recognize that the national outcomes they presumed would follow American policy relied on subnational or non-national interactions first.

A final example illustrates this paradox further. As it had through the 1960s, 1970s, and 1980s, Congress remained suspicious that American military assistance enabled dictators who violated their subjects' human rights.[171] At the same time, through the International Military Education and Training Program and its companion, the Informational Program, the State and Defense Departments reformulated those congressional concerns precisely into justifications for these exchange programs. "To our knowledge," Manolas and Samelson reported in 1990, "no other government which provides education and training to foreign militaries places an emphasis on human rights practices which even approaches the attention which the US attaches to this subject in its military schools."[172] Meanwhile, just a few minutes before Leahy paraded out the GAO report—which he personally commissioned—that criticized IMET's execution but upheld its "cost-effectiveness," he and Senator Dennis DeConcini (D-AZ) pressed undersecretary of state for international security affairs Reginald Bartholomew to explain a recent *Miami Herald* article describing widespread human rights abuses by fifteen El Salvadorian army officers—fourteen of whom were graduates of American IMET programs.[173] Bartholomew responded: "We have found, over the years, that one of the best tools we have for introducing foreign military personnel to American values is through the Individual [sic] Military Education and Training program."[174] He admitted, "The tragic situation in El Salvador is truly a hard test of our ability to encourage those values. Nonetheless, we remain convinced that our policy bolsters the center against violent extremists of the left and the right. IMET is a key component of that policy."

## Conclusion

Other historians have noted recently how the discourse of modernization has returned to American foreign policy in the last decade. Nils Gilman attributes its resurgence to the age of terrorism, where twenty-first-century Americans, seeking a national identity and national mission of comparable vigor to their "ideologically fanatical enemies," have unwittingly returned to the optimism and confidence of the 1950s.[175] Nick Cullather exposes the irony that the failed American-funded dam project in Afghanistan's Helmand Valley in the 1960s left the land useless for anything except poppies, which financed the Taliban's rise to power in the 1990s, and that the United States' reconstruction plan after the 2003 invasion included restoring the dam.[176] By these accounts, modernization theory reappeared after a well-earned absence.

In other places, modernization theory did not really go away after the Vietnam War; it continued in less conspicuous places. Programs that sponsored foreign military personnel in the United States continued for decades after the 1960s because they conformed to deeply held beliefs about the power of American expertise to transform the world. Quite the contrary, the programs were remarkably resilient and weathered the major shifts in American foreign policy in the 1960s, 1970s, and 1980s, despite the absence of evidence that they worked, because they appealed to a deeper faith that individual contact with ordinary Americans could inspire visitors to remake their societies. Because these programs yielded psychological, not physical, benefits, they were inexpensive. Since they depended so much on the initiative of local American communities for success, their practical execution and ongoing ideological sustenance flexibly adapted to major changes in American foreign policy during those decades.

American sponsor families for international military personnel probably would not have recognized their service as a project of American liberal internationalism, fueled by universalist ideologies of American exceptionalism and modernization theory. In 1959, the Draper Committee said, simply, that when foreign nationals lived in close contact with US citizens, they developed an appreciation for American values and way of life and that they went back home changed men and women.[177] This sentiment repeated, nearly verbatim, for decades, from senators, secretaries and undersecretaries of State and Defense, generals, and scholars of modernization, of realism, and of "new directions in foreign policy," although usually there was something more unsaid or implied about what that change entailed. Modernization theory absorbed the official policy for visiting military students, and it never went away completely. The ideological grooves of Military Internationalism in American communities preexisted and ran deeper than the careers of the great social scientists of the 1950s and 1960s.

The United States' ongoing military training exchange with Ghana in the years after the 1979 coup reveals how flexible and durable the rhetoric surrounding international military education could be. Instead of questioning the apparent lack of fruit of several decades of military exchanges, American policy makers aggressively defended international military education and training programs as a precarious toehold of American influence. In the minds of American planners, international military education preserved the ideal, Western norms of civil-military relations in the hearts of Ghanaian officers and, as Ghanaian politics continued to overwhelm the practicality of these ideals, American policy makers projected onto international military education the role of preserving the United States' entire diplomatic relationship with Ghana. Both of these missions—preserving the ideal of the professional, nonpolitical military and the ability of international military education and training programs to sustain American diplomatic relations with Ghana—vastly overestimated what it could actually accomplish in the short term. Neither seemed inappropriate, however, given the underlying faith that international military education and training exchanges still embodied the transformative potential of person-to-person contact between military officers from the Global South and the ideal American.[178] Since that faith never faltered during this time, the specific rhetoric that tapped into that faith to justify specific policies could be radically redirected without collapsing entirely.

# Conclusion

Between the 1960s and 1992, American security assistance planners made their program of sponsoring Ghanaian military members for training in the United States the only security assistance the United States provided Ghana. Roughly six hundred Ghanaian soldiers and their families attended training courses in the United States during that time.[1] Meanwhile, over thirty thousand Ghanaian soldiers participated in United Nations peacekeeping deployments in the Sinai and Lebanon over the same period. This book describes what change the elites that formed the American exchange program intended to inspire in the Ghanaian men and women who visited the United States and describes how the lives of Ghanaian families changed as a result of their military-sponsored international travel. It argues that the power to determine the meaning of global military exchanges flowed both ways, in both vertical and horizontal axes.

Vertically, both citizens and their governments appropriated the lofty official language that international military exchange transcended nations, transformed visitors into modern citizens, formed lasting relationships, and brought peace—when it suited them. Ghanaian soldiers and their families maximized the rewards of international education and training programs abroad or of UN peacekeeping service to exercise their individual strategies to mitigate Ghana's political and economic uncertainty. Government agents in Congress, the Department of Defense, and the State Department fell back on the unmeasurable human dimensions of international military training when forced to defend specific programs' lack of real results or, worse, evidence that American military training programs created monsters. Political and military elites in Ghana held up mythical, modern, military values of loyalty, professionalism, and nationalism they acquired via international sources to alternately discourage military intervention into Ghanaian politics, defend their intervention after the fact, or sustain peacekeeping operations that benefitted political and military leaders alike. Ghanaian citizens cited the new identity that international military travel exposed to make claims against their government. Governments also acknowledged that identity, if only to defend lucrative international exchange programs.

Horizontally, the internationalist identity that military exchange nurtured augmented the regional and national outlooks of individual Ghanaians and Americans. Ghanaian military families made the entire globe their workplace as they exercised economic strategies similar to other African families in the post-independence diaspora abroad. Meanwhile, both men and women capitalized on international circuits of hospitality to form long-lasting friendships around the world. The major difference was that a global military infrastructure—the United States' massive education and training program and the United Nations' unending mandate for peacekeeping deployments—both enabled and restricted individuals' freedom to travel. Likewise, individuals who attached themselves to a global military infrastructure while simultaneously affirming their commitment to peace ensured that the internationalist identity they professed was intellectually flexible enough to accommodate a wide range of seemingly contradictory values.

This work explored the history of the United States' international military education and training programs in the three decades after 1960 and Ghana's parallel participation in international training and education during that time. First, by evaluating the United States' international training program, it argued that American policy makers who thought they were contributing to the modernization of the "developing world" were actually restating prior assumptions about the transformative power of travel to the United States. They applied those assumptions to existing international military education and training exchange programs, and the programs became resilient enough to weather great changes in American foreign policy over three decades. Next, it examined how Ghanaian military planners and military-turned-political leaders consumed international military education assistance to educate Ghanaian officers, justify their own intervention into Ghanaian politics, and export military expertise to other African states. They cannily employed internationally recognized metaphors to explain their actions and plot their courses, but they also sincerely regarded some international military ethics and valued highly opportunities for international travel only accessible to military personnel. Policy makers in both the United States and Ghana assumed that international travel, especially for military elites or potential elites, could yield corporate transformation and modernization to recipient states' entire societies, via the military. But those advancements only occurred after *individual* transformations. The transformation to modernity they described was closer to a symptom of a global internationalist identity than the cause.

The individuals affected experienced the international exchange much more concretely. Ghanaian families on military-sponsored courses abroad also employed international education to exercise a global social imaginary based on entrepreneurial travel to relieve economic and political stresses in Ghana. Finally, large numbers of Ghanaian soldiers and their spouses integrated the trials and danger of international peacekeeping both for the benefits they provided and with a genuine faith that their service nurtured an authentically better world. These groups, in other words, saw international military travel as a method to access or actualize something better than the world of geopolitical competition between Cold War hegemons, the ravaging of their proxy wars, and the economic stagnation and neocolonialism that scorched the battlegrounds between them.

* * *

In 1993, three decades after deputy secretary of defense Roswell Gilpatrick initiated the "Informational Program for Foreign Military Trainees and Visitors in the United States," the undersecretary of defense for policy commissioned the RAND Corporation to evaluate once again whether training international military students delivered any measureable results.[2] RAND's Jennifer Morrison Taw and William H. McCoy repeated in 1993 many of the same conclusions that the State Department's Neil Campbell and Elliot Strauss made in 1971 and Harry Amos of the General Research Corporation reiterated in 1979: International students returned home with "positive impressions of the United States," yet there was "no proven means of evaluating the effectiveness of the IP." It was "virtually impossible" to assess "positive actions and non-actions" of US-trained personnel, but their "transgressions . . . such as the 1989 murder of Jesuit priests and two women by the Atlacatl battalion in El Salvador" were "often glaring."[3] In all likelihood, Taw and McCoy mused, post–Cold War budget cuts would render further discussion moot anyway, since "total US foreign military training is therefore declining inevitably."[4]

That 1993 concession was striking not because it was a fresh admission of failure atop a stack of likeminded reports three decades old, but because it still held out faith in the intangible benefits that flowed from personal relationships between international military students and Americans at all. In that manner, the Ghanaian case is representative universally: as an exercise in foreign policy, the United States' international military education and training program had no chance to mold a recipient state's national political development, democratic institutions, or commitment to human rights via intervention into the professional education of military leaders.

Scholars and government-funded studies alike repeated the same conclusion for decades, even when they continued to praise the program in principle. In places where the United States sought or had the capacity to maintain limited political influence, they considered even a token military exchange program worthwhile. Yet no one—neither policy makers in the United States or Ghana nor citizens of either country—considered the education and training of Ghanaian military personnel in the United States a failure. American political elites and citizens praised the program in principle because, as they ever were since the earliest days of American military exchange programs in the twentieth century, the specific sites of ideological transfer—American communities—were permanently above question. Likewise, Ghanaians celebrated the opportunity to travel to the United States for the individual personal, economic, and political benefits it brought. Military travel brought Ghanaians and Americans closer to the "imagined world," the interdependent global community. Even when graduates of American staff colleges overthrew their governments—as did Acheampong in 1972—no one claimed the relationships they formed with their sponsor families had anything to do with the coup.

Even elsewhere in Africa, where the United States ran small exchange programs, military education in the United States resulted in no noticeable national or corporate benefits on its own. RAND's William H. McCoy admitted that particularly in Africa, "where change and development have not occurred despite considerable investment by many countries around the world, training in the United States may simply be a respite for a deserving individual from an otherwise dreary existence."[5] US training brought no improvement to either the "professionalism or capabilities" of the Liberian army during the 1980s, for example, even though IMET had sponsored roughly the same number of Liberian military students as Ghanaians in the previous decade.[6] Ghana even helped lead the Economic Community of West African States' peacekeeping mission to Liberia in 1989. Meanwhile, recent scholarship suggests a causal connection between military men who receive American military education and training and their propensity to overthrow their governments.[7] At best, as the case of Senegal—where the United States trained roughly half as many soldiers as Ghanaians—showed, international military training was only one of several factors, including political stability, economic security, and ethnic harmony, that contributed to a state's civil-military relationship.[8] By demonstrating that this held true for Ghanaians as well, the case of Ghana is representative in that international military travel provided individual families opportunities to improve their own lives. Across the decades, Americans largely accepted this as a satisfactory outcome, if not the official one.

Compared to the United States' far larger military training initiatives in Southeast Asia in the 1960s and Latin America from the 1950s through the 1980s, the limited number of Ghanaian soldiers who trained in the United States makes the Ghanaian case unique. An international military education and training program like the one that brought small numbers of Ghanaian elites to the United States operated under considerably less menacing conditions. American military education and training that took place abroad needed other ways to reconcile the coercive aspects of modernization theory or Cold War anti-communism without the reserve of faith in the goodness of American communities to mask the programs' militarism.

Refusing to correlate the social imaginary of a global military community with graduates who return to terrorize their communities fails a critical test of internationalism. In Latin America or Southeast Asia, for example, a code of silence or refusal to hold fellow graduates accountable for human tragedies they caused appeared to be one corporate by-product of American military training. At least, highly publicized cases of American-trained soldiers torturing and murdering political opponents caused a great deal more soul-searching over whether the United States was simply training "assassins."[9] American training for those nations' armies occurred among American communities, to some extent, but far more often in home countries or in the Canal Zone. In the latter case, the internationalist faith that maintained the Informational Program never entered the picture. It is not clear that American training or life in American communities *caused* sadistic behavior. Describing Uruguay's dirty war of the mid-1970s, for example, *The New Yorker*'s Lawrence Weschler recalled that some officers insisted, "with an almost perverse hemispheric pride, [that] Latin American militaries were quite capable of coming up with all kinds of unpleasantness on their own, and needed no instruction from the North."[10] But more than 1,400 Uruguayans attended the School of the Americas in the Canal Zone between 1950 and 1977 when President Carter terminated all American military assistance to Uruguay. Therefore, the question of how American military training for soldiers in the client states also damaged the global community still lingers.[11]

That lingering question challenges the extent to which international military training and peacekeeping missions fit among other transnational paths to cooperation included in the definition of internationalism.[12] Without question, formal state machinery dominated the shape and structure of the exchanges and usually couched them in terms of geopolitical security.

On the other hand, both the individuals who traveled and the individuals whose hospitality nurtured relationships with the travelers devised their own strategies to extract meaning from that travel, be it economic, political, or cultural. Even policy makers, when pressed, admitted that those individual meanings—relationships "of a higher order," a peacekeeper with "something up my sleeve to surprise the world with"—were more than just a cost-free benefit; they were closer to the "real world" after all.

# Notes

## FOREWORD

1. Thomas J. Noer, "New Frontiers and Old Priorities in Africa," in *Kennedy's Quest for Victory*, ed. Thomas G. Paterson, (New York: Oxford University Press, 1989). Also see from Noer, *Cold War and Black Liberation: The United States and White Rule in Africa, 1948–1968* (Columbia: University of Missouri Press, 1985) and "The New Frontier and African Neutralism: Kennedy, Nkrumah, and the Volta River Project" *Diplomatic History* 8, no. 1 (1984): 61–79; Thomas Borstelmann, *Apartheid's Reluctant Uncle: The United States and Southern Africa in the Early Cold War* (New York: Oxford University Press, 1993) and *Cold War and the Color Line: American Race Relations in the Global Arena* (Cambridge, MA: Harvard University Press, 2001); Richard Mahoney, *JFK: Ordeal in Africa* (New York: Oxford University Press, 1983); Piero Gleijeses, *Conflicting Missions: Havana, Washington, and Africa, 1959–1976* (Chapel Hill: University of North Carolina Press, 2002); Andrew DeRoche, *Black, White, and Chrome: The United States and Zimbabwe, 1953 to 1998* (Trenton, NJ: Africa World Press, 2001).
2. See Ryan Irwin, *Gordian Knot: Apartheid and the Unmaking of the Liberal World Order* (New York: Oxford University Press, 2012); William L. Bishop, "Diplomacy in Black and White: America and the Search for Zimbabwean Independence, 1965–1980" (PhD Dissertation, Vanderbilt University, 2012); Jamie Miller, "South Africa in the Cold War, 1974–6" (PhD Dissertation, Cambridge University, 2014); Philip E. Muehlenbeck, *Betting on the Africans: John F. Kennedy's Courting of African Nationalist Leaders* (New York: Oxford University Press, 2012).
3. Frank Kramer, *Foreign Assistance and Related Programs Appropriations for 1986*, 99th Cong., 1st Sess. (March 26, 1985), 517–18. Technically, the IMET program did not formally exist before 1976, but both policy makers and scholars used the term International Military Education and Training

and the acronym IMET before then. This book only uses the acronym when referring to the specific, formal program.

## INTRODUCTION

1. Kwame Anthony Appiah, *Cosmopolitanism: Ethics in a World of Strangers*, Issues of Our Time (New York: W. W. Norton, 2006).
2. See, for example, Bradley R. Simpson, *Economists with Guns: Authoritarian Development and U.S.-Indonesian Relations, 1960–1968* (Stanford, CA: Stanford University Press, 2008); Nick Cullather, *The Hungry World: America's Cold War Battle Against Poverty in Asia* (Cambridge, MA: Harvard University Press, 2010); Jeffrey Taffet, *Foreign Aid As Foreign Policy: The Alliance for Progress in Latin America* (New York: Routledge, 2007); Thomas C. Field, "Ideology as Strategy: Military-Led Modernization and the Origins of the Alliance for Progress in Bolivia," *Diplomatic History* 36, no. 1 (January 1, 2012): 147–83; Robert H. Holden, "The Real Diplomacy of Violence: United States Military Power in Central America, 1950–1990," *International History Review* 15, no. 2 (1993): 283–322; John S. Fitch, "The Political Impact of U.S. Military Aid to Latin America: Institutional and Individual Effects," *Armed Forces and Society* 5, no. 3 (Spring 1979): 360–86; Chester A. Crocker, "Military Dependence: The Colonial Legacy in Africa," *Journal of Modern African Studies* 12, no. 2 (June 1, 1974): 265–86; James A. Bill, "The Military and Modernization in the Middle East," *Comparative Politics* 2, no. 1 (October 1969): 41.
3. Odd Arne Westad, *The Global Cold War: Third World Interventions and the Making of Our Times* (Cambridge: Cambridge University Press, 2007), 5; Melvyn P. Leffler, *For the Soul of Mankind: The United States, the Soviet Union, and the Cold War* (New York: Hill and Wang, 2007), 8.
4. Philip E. Muehlenbeck, *Betting on the Africans: John F. Kennedy's Courting of African Nationalist Leaders* (Oxford: Oxford University Press, 2012). See David E. Apter, *The Gold Coast in Transition* (Princeton, NJ: Princeton University Press, 1955); David E. Apter, *Ghana in Transition*, 2nd rev. ed. (Princeton, NJ: Princeton University Press, 1972); Elizabeth Cobbs Hoffman, *All You Need Is Love: The Peace Corps and the Spirit of the 1960s* (Cambridge, MA: Harvard University Press, 1998), 157; Gabriel Kolko, *Confronting the Third World: United States Foreign Policy, 1945–1980* (New York: Pantheon Books, 1988), 198.
5. James H. Meriwether, "'Worth a Lot of Negro Votes': Black Voters, Africa, and the 1960 Presidential Campaign," *Journal of American History* 95, no. 3 (December 1, 2008): 737–63; Muehlenbeck, *Betting on the Africans*;

see also Mary L. Dudziak, *Cold War Civil Rights: Race and the Image of American Democracy* (Princeton, NJ: Princeton University Press, 2000); Kevin Kelly Gaines, *American Africans in Ghana: Black Expatriates and the Civil Rights Era* (Chapel Hill: University of North Carolina Press, 2006), 5; James T. Campbell, *Middle Passages: African American Journeys to Africa, 1787–2005* (New York: Penguin Press, 2006); Jason C. Parker, "'Made-in-America Revolutions'? The 'Black University' and the American Role in the Decolonization of the Black Atlantic," *Journal of American History* 96, no. 3 (December 1, 2009): 727–50; Horace Campbell, "Pan-Africanism and African Liberation," in *Imagining Home: Class, Culture, and Nationalism in the African Diaspora*, eds. Sidney J. Lemelle and Robin D. G. Kelley (London: Verso, 1994), 285–307; Penny M. Von Eschen, *Race Against Empire: Black Americans and Anticolonialism, 1937–1957* (Ithaca, NY: Cornell University Press, 1997).

6. See, for example, Michael Hunt, *Ideology and U.S. Foreign Policy* (New Haven, CT: Yale University Press, 2009), 3.
7. See Muehlenbeck, *Betting on the Africans*; Gaines, *American Africans in Ghana*.
8. Robert B. Rakove, *Kennedy, Johnson, and the Nonaligned World* (Cambridge: Cambridge University Press, 2013), 138; John F. Kennedy, "242. Letter from President Kennedy to President Nkrumah; December 14, 1961," in *FRUS 1961–1963*, vol. xxi, *Africa*, ed. Nina Davis Howland (Washington, DC: US Govt. Print. Off., 1995), 371–72; Westad, *The Global Cold War*, 2, 98, 107; Hoffman, *All You Need Is Love*, 154.
9. Kwame Nkrumah, *Revolutionary Path* (London: Panaf, 1973), 173; "232. Memorandum of Telephone Conversation between President Kennedy and the Acting Under Secretary of State (Ball); September 21, 1961," in *FRUS 1961–1963*, vol. xxi, *Africa*, ed. Nina Davis Howland (Washington, DC: US Govt. Print. Off., 1995), 355; Kennedy, "242. Letter from President Kennedy to President Nkrumah; December 14, 1961," 371–72; Hoffman, *All You Need Is Love*, 154.
10. Rakove, *Kennedy, Johnson, and the Nonaligned World*, xx–xxi.
11. Ibid., xxi.
12. Ibid.; Philip E. Muehlenbeck, "Kennedy and Touré: A Success in Personal Diplomacy," *Diplomacy and Statecraft* 19 (2008): 69–95; Muehlenbeck, *Betting on the Africans*.
13. Hoffman, *All You Need Is Love*, 157; Muehlenbeck, *Betting on the Africans*, 77; McGeorge Bundy, "240. Notes for the Record; December 5, 1961; Subject: NSC Meeting on Volta Dam," in *FRUS 1961–1963*, vol. xxi,

*Africa*, ed. Nina Davis Howland (Washington, DC: US Govt. Print. Off., 1995), 369; See McGeorge Bundy, "239. Memorandum from the President's Special Assistant for National Security Affairs (Bundy) to President Kennedy; Subject: The Volta Project; December 1, 1961," in *FRUS 1961–1963*, vol. xxi, *Africa*, ed. Nina Davis Howland (Washington, DC: US Gov. Print. Off., 1995), 365–68; Rakove, *Kennedy, Johnson, and the Nonaligned World*, 64, 138.

14. Westad, *The Global Cold War*, 396.
15. Ibid., 107.
16. Hoffman, *All You Need Is Love*, 157; Bundy, "240. Notes for the Record; December 5, 1961; Subject: NSC Meeting on Volta Dam," 369. See Bundy, "239. Bundy to Kennedy; December 1, 1961; Subj: Volta Project."
17. Bundy, "239. Bundy to Kennedy; December 1, 1961; Subj: Volta Project"; Lloyd Garrison, "Nkrumah Seeks U.S. Aid Decision: Asks Action on Ghana River Project by Oct. 13," *New York Times*, October 3, 1961; "232. Kennedy to Ball; September 21, 1961," 355; Elizabeth A. Cobbs, "Decolonization, the Cold War, and the Foreign Policy of the Peace Corps," *Diplomatic History* 20, no. 1 (1996): 79; Michael E. Latham, *Modernization as Ideology: American Social Science and "Nation Building" in the Kennedy Era* (Chapel Hill: University of North Carolina Press, 2000), 8.
18. Latham, *Modernization as Ideology*, ix.
19. Michael Adas, *Machines as the Measure of Men: Science, Technology, and Ideologies of Western Dominance* (Ithaca, NY: Cornell University Press, 1989); Michael Adas, *Dominance by Design: Technological Imperatives and America's Civilizing Mission* (Cambridge, MA: Belknap Press of Harvard University Press, 2006); David Ekbladh, *The Great American Mission: Modernization and the Construction of an American World Order* (Princeton, NJ: Princeton University Press, 2009).
20. Thomas J. McCormick, *America's Half-Century: United States Foreign Policy in the Cold War and After* (Baltimore: Johns Hopkins University Press, 1995), chap. 4; George F. Kennan, *American Diplomacy* (Chicago: University of Chicago Press, 1984); Walt W. Rostow, *The Process of Economic Growth: A Non-Communist Manifesto*, 2nd ed. (Oxford: Clarendon Press, 1960); James Cronin, *The World the Cold War Made: Order, Chaos, and the Return of History* (New York: Routledge, 1996); Westad, *The Global Cold War*, 24. On whether the United States really had the world's best interests in mind, see William Appleman Williams, *The Tragedy of American Diplomacy* (New York: Norton, 1988); Kolko, *Confronting the Third World*.
21. Daniel Lerner, *The Passing of Traditional Society: Modernizing the Middle East* (Glencoe, IL: Free Press, 1958). Among the most influential: Max F. Millikan

and Walt W. Rostow, *A Proposal: Key to an Effective Foreign Policy* (New York: Harper & Brothers, 1957); Lucian W. Pye, *Guerrilla Communism in Malaya: Its Social and Political Meaning* (Princeton, NJ: Princeton University Press, 1956); Rostow, *The Process of Economic Growth: A Non-Communist Manifesto*; Nick Cullather, "Modernization Theory," in *Explaining the History of American Foreign Relations,* eds. Michael J. Hogan and Thomas G. Paterson, 2nd ed. (Cambridge: Cambridge University Press, 2004), 212–20. See also, Nils Gilman, *Mandarins of the Future: Modernization Theory in Cold War America* (Baltimore: Johns Hopkins University Press, 2003); Latham, *Modernization as Ideology.*

22. Latham, *Modernization as Ideology.*
23. By 1962, 31 of the UN General Assembly's 110 members were from Africa. Muehlenbeck, *Betting on the Africans*, xi.
24. Frederick Cooper, *Colonialism in Question: Theory, Knowledge, History* (Berkeley: University of California Press, 2005), 113.
25. Ibid.
26. See Kwaku Larbi Korang, *Writing Ghana, Imagining Africa: Nation and African Modernity*, Rochester Studies in African History and the Diaspora (Rochester, NY: University of Rochester Press, 2003), 5.
27. Peter Braham, *Key Concepts in Sociology*, SAGE Key Concepts (London: SAGE, 2013), 7.
28. "The Charter of the Ghana Military Academy; Appx II to Annex D to GA/569/1/0 (SD); August 26, 1959" (Accra, Ghana, August 26, 1959), 3, RG 14/4/504 Ghana Military Academy, doc. Formation of Ghana Army Training Establishment, PRAAD.
29. See introduction to Chapter 2. Francis K. Russell, "News Release by United States Embassy, Accra; September 11, 1961; Subject: Remarks of the American Ambassador Mr. Francis K. Russell at the Presentation of Books to the Ghana Military Academy" (Burma Camp, Accra, September 11, 1961), RG 14/4/506 Ghana Military Academy, doc. 201, PRAAD.
30. Benedict Anderson, *Imagined Communities: Reflections on the Origin and Spread of Nationalism*, rev. ed. (London: Verso, 2006), 116–126.
31. Lt. Col. C. R. Tachie-Menson, "Education of Military Leaders," *Ghana Armed Forces Magazine*, October 1969, 12, Bibliothek der Friedrich-Ebert-Stiftung, Bonn, Germany (Hereafter: FES); Samuel P. Huntington, *The Soldier and the State: The Theory and Politics of Civil-Military Relations* (Cambridge, MA: Belknap Press of Harvard University Press, 1957); Samuel P. Huntington, *Changing Patterns of Military Politics*, vol. 3, International Yearbook of Political Behavior Research (New York: Free Press of Glencoe, 1962); Morroe Berger, *Military Elite and Social Change* (Princeton, NJ:

Center for International Studies, Princeton University, 1960); John J. Johnson, ed., *The Role of the Military in Underdeveloped Countries* (Princeton, NJ: Princeton University Press, 1962); William F. Gutteridge, *Military Institutions and Power in the New States* (New York: Praeger, 1965).

32. Rostow, *The Process of Economic Growth*, 7, 59, 4.
33. Braham, *Key Concepts in Sociology*, 7–8; Edward Shils, "The Military in the Political Development of the New States," in *The Role of the Military in Underdeveloped Countries*, ed. John J. Johnson (Princeton, NJ: Princeton University Press, 1962), 10–11; Gilman, *Mandarins of the Future*, 12, 146; Chandan Reddy, "Modern," in *Keywords for American Cultural Studies*, eds. Bruce Burgett and Glenn Hendler (New York: New York University Press, 2007), 161.
34. Samuel P. Huntington, *Political Order in Changing Societies* (New Haven, CT: Yale University Press, 1973), 193; Harry O. Amos et al., *U.S. Training of Foreign Military Personnel, Volume 2: Main Report* (McLean, VA: General Research Corporation, Tactical Warfare Operations, March 1979), 13; Richard F. Grimmett, *Foreign Assistance and Related Programs Appropriations for 1986* (Washington, DC: US Govt. Print. Off., 1985), 432; Bruce E. Arlinghaus, *Military Development in Africa: The Political and Economic Risks of Arms Transfers* (Boulder, CO: Westview Press, 1984), 8; J. Bayo Adekson, "Pay, Promotion, and Other Self-Interests of Military Intervention in Politics," *Military Affairs* 45, no. 1 (February 1, 1981): 21; Robert A. Packenham, *Liberal America and the Third World: Political Development Ideas in Foreign Aid and Social Science* (Princeton, NJ: Princeton University Press, 1973), 228; Samuel Decalo, "Military Rule in Africa: Etiology and Morphology," in *Military Power and Politics in Black Africa*, ed. Simon Baynham (London: Croom Helm, 1986), 41.
35. Gilman, *Mandarins of the Future*; Ekbladh, *The Great American Mission*; Cullather, *The Hungry World*; Cullather, "Modernization Theory"; Latham, *Modernization as Ideology*.
36. Lerner, *The Passing of Traditional Society*; Millikan and Rostow, *A Proposal*; Rostow, *The Process of Economic Growth*; Pye, *Guerrilla Communism in Malaya*.
37. Cooper, *Colonialism in Question*, chap. 5.
38. C. A. Bayly, *The Birth of the Modern World: 1780–1914* (Malden, MA: Blackwell, 2004), 10.
39. Ibid., 11; Reddy, "Modern," 161; Braham, *Key Concepts in Sociology*, 7–8; Cooper, *Colonialism in Question*, 131.
40. Cooper, *Colonialism in Question*, 113, 131.

41. Ibid., 131.
42. Korang, *Writing Ghana, Imagining Africa*, 2–13; Michael Amoah, *Reconstructing the Nation in Africa: The Politics of Nationalism in Ghana*, International Library of African Studies 19 (London: Palgrave Macmillan, 2007), 5.
43. Anderson, *Imagined Communities*, 120, n. 9.
44. Kwame Anthony Appiah, *In My Father's House: Africa in the Philosophy of Culture* (New York: Oxford University Press, 1992), 162.
45. Emmanuel Akyeampong, "'Diasporas,' Mobility and the Social Imaginary: Getting Ahead in West Africa," *Journal of Third World Studies* 27, no. 1 (Spring 2010): 25–41; Simon Baynham, *The Military and Politics in Nkrumah's Ghana* (Boulder, CO: Westview Press, 1988); Eboe Hutchful, "Institutional Decomposition and Junior Ranks' Political Action in Ghana," in *The Military and Militarism in Africa*, eds. Eboe Hutchful and Abdoulaye Bathily (Dakar: Codesria, 1998), 211–56; Eboe Hutchful, "Reconstructing Civil-Military Relations and the Collapse of Democracy in Ghana, 1979–81," *African Affairs* 96, no. 385 (October 1, 1997): 535–60.
46. Emmanuel Akyeampong, "Africans in the Diaspora: The Diaspora and Africa," *African Affairs* 99, no. 395 (April 1, 2000): 183–215; Akyeampong, "'Diasporas,' Mobility and the Social Imaginary"; Amoah, *Reconstructing the Nation in Africa*; John A. Arthur, *African Diaspora Identities: Negotiating Culture in Transnational Migration* (Lanham, MD: Lexington Books, 2010); Leif O. Manger and Munzoul A. M. Assal, "Diasporas Within and Without Africa—Dynamism, Heterogeneity, Variation," in *Diasporas Within and Without Africa: Dynamism, Heterogeneity, Variation*, eds. Lief O. Manger and Munzoul A. M. Assal (Uppsala, Sweden: Nordiska Afrikainstitutet, 2006), 7–31; Takyiwaa Manuh, "'Efie' or the Meanings of 'Home' Among Female and Male Ghanaian Migrants in Toronto, Canada and Returned Migrants to Ghana," in *New African Diasporas*, ed. Khalid Koser (London: Routledge, 2003), 140–59; Paul Tiyambe Zeleza, "Reconceptualizing African Diasporas: Notes from a Historian," *Transforming Anthropology* 18, no. 1 (2010): 74–78.
47. Thomas W. Zeiler, "Just Do It! Globalization for Diplomatic Historians," *Diplomatic History* 25, no. 4 (2001): 531. See also, Micol Seigel, "Beyond Compare: Comparative Method after the Transnational Turn," *Radical History Review* 2005, no. 91 (December 21, 2005): 62–90; Micol Seigel, *Uneven Encounters: Making Race and Nation in Brazil and the United States*, American Encounters/Global Interactions Series (Durham, NC: Duke University Press, 2009), xii; Jose C. Moya, *Cousins and Strangers: Spanish Immigrants in Buenos Aires, 1850–1930* (Berkeley: University of California Press, 1998), 4.

48. Akira Iriye, *Cultural Internationalism and World Order* (Baltimore: Johns Hopkins University Press, 1997), 2–3.
49. Mazower suggests the blend of international organizations, trans- or sub-national organizations, and the metastructures of international cooperation might better be called "intergovernmentalism." Mark Mazower, *Governing the World: The History of an Idea* (New York: Penguin Press, 2012), xiii.
50. Akira Iriye, "Internationalizing International History," in *Rethinking American History in a Global Age*, ed. Thomas Bender (Berkeley: University of California Press, 2002), 53.
51. "Political development" and "professionalism" were two of Huntington's most influential contributions to the scholarship of civil-military relations in the 1960s and 1970s. Samuel P. Huntington, "Political Development and Political Decay," *World Politics* 17, no. 3 (April 1, 1965): 386–430; Huntington, *The Soldier and the State*, chap. 4; Huntington, *Political Order in Changing Societies*.
52. Minus "military travel," this is Iriye's definition of "cultural internationalism." Iriye, *Cultural Internationalism and World Order*, 2.
53. Festus B. Aboagye, *The Ghana Army: A Concise Contemporary Guide to Its Centennial Regimental History, 1897–1999* (Accra, Ghana: Sedco, 1999), 119; S. Kojo Addae, *History of Ghana Armed Forces: Military Organizations and Regiments*, vol. 3, *Military Histories* (Accra, Ghana: Eureka Publications, c2010), 3.
54. For a similar justification, see Catherine Lutz, *Homefront: A Military City and the American Twentieth Century* (Boston: Beacon Press, 2001), 1, n. 1.
55. David Wiley, "Militarizing Africa and African Studies and the U.S. Africanist Response," *African Studies Review* 55, no. 2 (2012): 147–61.
56. See, for example, Akyeampong, "'Diasporas,' Mobility and the Social Imaginary"; Manuh, "'Efie' or the Meanings of 'Home'"; Thomas Y. Owusu, "The Role of Ghanaian Immigrant Associations in Toronto, Canada," *International Migration Review* 34, no. 4 (December 1, 2000): 1155–81.

## CHAPTER 1

1. John F. Kennedy, "Statement by the President upon Signing the Foreign Assistance Act—September 4, 1961," *Public Papers of the Presidents of the United States* 1961 (September 4, 1961): 588.
2. John F. Kennedy, "Special Message to the Congress on Foreign Aid—March 22, 1961," *Public Papers of the Presidents of the United States* 1961 (March 22, 1961), 205; John F. Kennedy, "Urgent National Needs: Address of the President of the United States," Doc. No. 87-174, at 2 (May 25, 1961).
3. Kennedy, "Special Message to the Congress on Foreign Aid—March 22, 1961," 203.

4. John F. Kennedy, "Statement by the President on Ordering Resumption of Underground Nuclear Tests—September 5, 1961," *Public Papers of the Presidents of the United States* 1961 (September 5, 1961), 589.
5. John F. Kennedy, "National Security Action Memorandum No. 88; Memorandum to the Secretary of Defense; Subject: Training for Latin American Armed Forces," September 5, 1961, Federation of American Scientists Intelligence Resource Program, *www.fas.org/irp/offdocs/nsam-jfk/index.html.*
6. Ibid.
7. Ibid.
8. John F. Kennedy, "National Security Action Memorandum No. 89; Memorandum to the Secretary of State; Subject: Soviet Training of Ghanaian Troops," September 5, 1961, Federation of American Scientists Intelligence Resource Program, *www.fas.org/irp/offdocs/nsam-jfk/index.html.* Specifically, Nkrumah proposed to send officer cadets to the Soviet Union, against the advice of his British expatriate chief of defense staff (who was sacked later that month).
9. Roger Hilsman Jr., "231. Memorandum from the Director of the Bureau of Intelligence and Research (Hilsman) to Secretary of State Rusk; Intelligence Note; Internal Dissention in Ghana; September 14, 1961," in *FRUS 1961–1963*, vol. xxi, *Africa*, ed. Nina Davis Howland (Washington, DC: US Govt. Print. Off., 1995), 354; William C. Trimble, "237. Memorandum from the Director of the Office of West African Affairs (Trimble) to the Assistant Secretary of State for African Affairs (Williams), February 11, 1964," in *FRUS 1964–1968*, vol. xxiv, *Africa*, ed. Nina Davis Howland (Washington, DC: US Govt. Print. Off., 1999), 413–16; "238. Special National Intelligence Estimate; SNIE 64.1-61; PROSPECTS FOR GHANA; November 16, 1961," in *FRUS 1961–1963*, vol. xxi, *Africa*, ed. Nina Davis Howland (Washington, DC: US Govt. Print. Off., 1995), 363–64. See, for example, Bundy, "239. Bundy to Kennedy; December 1, 1961; Subj: Volta Project"; Bundy, "240. Notes for the Record; December 5, 1961; Subject: NSC Meeting on Volta Dam." See also Gaines, *American Africans in Ghana*, 5–24; Hoffman, *All You Need Is Love*, 157.
10. Quoted in Dana Adams Schmidt, "Nkrumah Confers at White House; Backs U.N. Effort," *New York Times*, March 9, 1961, ProQuest Historical Newspapers.
11. Cobbs, "Decolonization, the Cold War"; Hoffman, *All You Need Is Love*, 151.
12. "232. Kennedy to Ball; September 21, 1961," 355.
13. John F. Kennedy, "National Security Action Memorandum No. 119; Memorandum to the Secretary of State and Secretary of Defense; Subject:

Civic Action," December 18, 1961, Federation of American Scientists Intelligence Resource Program, *www.fas.org/irp/offdocs/nsam-jfk/index.html*; Daniel Lerner and Richard D. Robinson, "Swords and Ploughshares: The Turkish Army as a Modernizing Force," *World Politics* 13, no. 1 (October 1, 1960): 34; Charles Windle and T. R. Vallance, "Optimizing Military Assistance Training," *World Politics* 15, no. 1 (October 1, 1962): 94.

14. See also, Ekbladh, *The Great American Mission*; Ian Tyrrell, *Reforming the World: The Creation of America's Moral Empire* (Princeton, NJ: Princeton University Press, 2010); Mazower, *Governing the World*; Gilman, *Mandarins of the Future*; Cullather, *The Hungry World*.
15. Cullather, "Modernization Theory," 213–217; Latham, *Modernization as Ideology*, 5, 55.
16. Lerner, *The Passing of Traditional Society*; Rostow, *The Process of Economic Growth*.
17. Walt W. Rostow, *The Diffusion of Power: An Essay in Recent History* (New York: Macmillan, 1972); Lerner and Robinson, "Swords and Ploughshares"; Walt W. Rostow, "94. Memorandum from the President's Deputy Special Assistant for National Security Affairs (Rostow) to President Kennedy; February 28, 1961; Subject: Crucial Issues in Foreign Aid," in *FRUS 1961–1963*, vol. ix, *Foreign Economic Policy*, ed. Evans Gerakas et al. (Washington, DC: US Govt. Print. Off., 1995), 204–9; Gilman, *Mandarins of the Future*, 50.
18. Gilman, *Mandarins of the Future*, 50; Westad, *The Global Cold War*.
19. Latham, *Modernization as Ideology*; Cullather, *The Hungry World*; Ekbladh, *The Great American Mission*, chap. 6; Simpson, *Economists with Guns*.
20. Cullather, *The Hungry World*, 109, 112; Field, "Ideology as Strategy"; Gilman, *Mandarins of the Future*, 11, 50.
21. Ernest W. Lefever, "The Military Assistance Training Program," *Annals of the American Academy of Political and Social Science* 424 (March 1, 1976): 85–95.
22. Amos et al., *U.S. Training of Foreign Military Personnel, Volume 2*, 101.
23. Mazower, *Governing the World*, 295; Noam Chomsky, *American Power and the New Mandarins* (New York: Pantheon Books, 1969); Chomsky refers to Ithiel De Sola Pool, "The Necessity for Social Scientists Doing Research for Governments," *Background* 10, no. 2 (August 1, 1966): 111; Gilman refers to both Pool and Chomsky in *Mandarins of the Future*, 8.
24. See also Hoffman, *All You Need Is Love*; Latham, *Modernization as Ideology*.
25. Jeremi Suri, *Power and Protest: Global Revolution and the Rise of Détente* (Cambridge, MA: Harvard University Press, 2003), chap. 4.
26. Chester J. Pach, *Arming the Free World: The Origins of the United States Military Assistance Program, 1945–1950* (Chapel Hill: University of North

Carolina Press, 1991), 198; Shils, "The Military in the Political Development of the New States," 86.

27. David S. Painter, "Explaining U.S. Relations with the Third World," *Diplomatic History* 19, no. 3 (Summer 1995): 525; Westad, *The Global Cold War*, 21.
28. This is Leffler's argument in *A Preponderance of Power: National Security, the Truman Administration and the Cold War* (Stanford, CA: Stanford University Press, 1992). For Leffler's more recent, optimistic contention that leaders in both Moscow and Washington believed they "possessed the formula for the good life," see *For the Soul of Mankind*, 8; Richard F Grimmett, *An Overview of United States Military Assistance Programs*, CRS Report for Congress (Washington, DC: Congressional Research Service, March 29, 1988), CRS—1; H. Comm. on Government Operations, Use of Defense Support Funds for Economic and Political Purposes, Twenty-First Report by the Committee on Government Operations, H.R. Rep. No. 85-1374, at 11 (1958); A Report on United States Foreign Operations by Honorable Allen J. Ellender, United States Senator from the State of Louisiana. S. Rep. No. 87-20 (March 15, 1961); John F. Kennedy, "Report to the Congress on the Foreign Assistance Program for the Fiscal Year 1962," Serial Set Vol. No. 12611, Session Vol. No. 47 (Washington, DC: 88th Cong., 1st Sess., August 28, 1963).
29. Dwight D. Eisenhower, "Report to Congress on the Mutual Security Program for the First Half of Fiscal Year 1960," May 2, 1960, 2.
30. Windle and Vallance, "Optimizing Military Assistance Training," 91.
31. Kennedy, "Urgent National Needs"; Kennedy, "Special Message to the Congress on Foreign Aid—March 22, 1961"; Rostow, "94. Rostow to Kennedy; February 28. 1961; Subj: Crucial Issues in Foreign Aid," 207.
32. Kennedy, "Urgent National Needs."
33. Technical Assistance was the fourth point on the list of development initiatives Truman spelled out in his second inaugural address, earning the generic title Point Four. Harry S. Truman, "Inaugural Address—Thursday, January 20, 1949," *Inaugural Addresses of the Presidents of the United States* 1989, 289; Leffler, *A Preponderance of Power*, 266.
34. Vernon W. Ruttan, *United States Development Assistance Policy: The Domestic Politics of Foreign Economic Aid*, Johns Hopkins Studies in Development Series (Baltimore: Johns Hopkins University Press, 1996); See Leffler, *A Preponderance of Power*, 10–15.
35. See Sen. Wiley's comments, p. 51 of *International Development and Security. Hearings Before the Committee on Foreign Relations, United States Senate, on S.*

*1983. 87th Cong., 1st Sess.,* 2 vols. (1961); Kenneth Alan Osgood, *Total Cold War: Eisenhower's Secret Propaganda Battle at Home and Abroad* (Lawrence: University of Kansas Press, 2006). Osgood contends that Eisenhower contrived all cultural diplomacy, but others give more credence to its success. See, for example, Reinhold Wagnleitner, "The Empire of the Fun, or Talkin' Soviet Union Blues: The Sound of Freedom and U.S. Cultural Hegemony in Europe," *Diplomatic History* 23, no. 3 (1999): 499–524; Penny M. Von Eschen, *Satchmo Blows up the World: Jazz Ambassadors Play the Cold War* (Cambridge, MA: Harvard University Press, 2004); Emily S. Rosenberg, *Spreading the American Dream* (New York: Hill and Wang, 1982), chap. 10; Rob Kroes, "American Empire and Cultural Imperialism: A View from the Receiving End," *Diplomatic History* 23, no. 3 (1999): 463–77.

36. Von Eschen, *Satchmo Blows up the World,* 6; Dwight D. Eisenhower, "Remarks at the People-to-People Conference—September 11, 1956," *Public Papers of the Presidents of the United States* 1956 (1956), 750; See also Akira Iriye, *Global Community the Role of International Organizations in the Making of the Contemporary World* (Berkeley: University of California Press, 2002), chap. 3.
37. H. W. Brands, "The Age of Vulnerability: Eisenhower and the National Insecurity State," *American Historical Review* 94, no. 4 (1989): 963–89, argues that this was as much a symptom of Eisenhower's leadership style as the consequence of the growth of bureaucracy.
38. Eisenhower, "Report to Congress on the Mutual Security Program for the First Half of Fiscal Year 1960," 2.
39. Bryce N. Harlow, "214. Memorandum for the Record by the President's Assistant (Harlow); January 30, 1958," in *FRUS 1958–1960,* vol. iv, *Foreign Economic Policy,* ed. Suzanne E. Coffman, Edward C. Keefer, and Harriet Dashiell Schwar (Washington, DC: US Govt. Print. Off., 1992), 412.
40. Ibid., 412, 409; "239. Memorandum of Conference with President Eisenhower; April 23, 1959," in *FRUS 1958–1960,* vol. iv, *Foreign Economic Policy,* ed. Suzanne E. Coffman, Edward C. Keefer, and Harriet Dashiell Schwar (Washington, DC: U.S. Govt. Print. Off., 1992), 454; "238. Notes of the Legislative Leadership Meeting; April 22, 1959," in *FRUS 1958–1960,* vol. iv, *Foreign Economic Policy,* ed. Suzanne E. Coffman, Edward C. Keefer, and Harriet Dashiell Schwar (Washington, DC: US Govt. Print. Off., 1992), 451–52.
41. Eisenhower, "Report to Congress on the Mutual Security Program for the First Half of Fiscal Year 1960," 6.
42. A Report on United States Foreign Operations by Honorable Allen J. Ellender, at 1.

43. H. Comm., on Government Operations, Use of Defense Support Funds for Economic and Political Purposes, at 12. Emphasis in original.
44. See, for example, A Report on United States Foreign Operations by Honorable Allen J. Ellender; Report of the Committee on Foreign Relations on S. 1451, Together with Minority Views, S. Rep. No. 86-412 (June 22, 1959); Report of the Committee on Foreign Relations, United States Senate, on S. 1983, S. Rep. No. 87-612 (July 24, 1961); H. Comm. on Foreign Affairs, Report of the Committee on Foreign Affairs on H.R. 11510 to Amend Further the Mutual Security Act of 1954, H.R. Rep. No. 86-1464 (April 7, 1960); H. Comm. on Foreign Affairs, Report of the Special Study Mission to Asia, Western Pacific, Middle East, Southern Europe, and North Africa, H.R. Rep. No. 86-1386 (March 14, 1960); H. Comm. on Government Operations, Use of Defense Support Funds for Economic and Political Purposes; John F. Kennedy, "Letter to the President of the Senate and to the Speaker of the House Transmitting Bill Implementing the Message on Foreign Aid—May 26, 1961," *Public Papers of the Presidents of the United States* 1961 (1961), 407; Michael J. Mansfield, "Comments to Senate," in *Congressional Record—Senate*, vol. 107 (1961), 16400–404. See Sen. Morse's comments in *Hearings Before the Senate Committee On Foreign Relations on the Foreign Assistance Act of 1961*, 65; Ruttan, *United States Development Assistance Policy*; Cronin, *The World the Cold War Made*.
45. US President's Committee to Study the US Military Assistance Program (1958–1959), *Conclusions Concerning the Mutual Security Program. Communication from the President of the United States Transmitting the Final Report of the President's Committee to Study the United States Military Assistance Program, with the Several Studies Which Are Annexes Thereto. August 20, 1959*," H.R. Doc. No. 86-215, part 1 at 22 (1959).
46. Report of the Committee on Foreign Relations, United States Senate, on S. 1983, at 4.
47. Ibid.
48. John F. Dulles, "226. Memorandum from Secretary of State Dulles to President Eisenhower; September 13, 1958," in *FRUS 1958–1960*, vol. iv, *Foreign Economic Policy*, ed. Suzanne E. Coffman, Edward C. Keefer, and Harriet Dashiell Schwar (Washington, DC: US Govt. Print. Off., 1992), 429–30.
49. Ibid.
50. "Our History," Draper Investment Company, accessed October 30, 2013, *www.draperco.com/english/our-history/*.
51. David L. Snead, *The Gaither Committee, Eisenhower, and the Cold War* (Columbus: Ohio State University Press, 1999).

52. See Morton H. Halperin, "The Gaither Committee and the Policy Process," *World Politics* 13, no. 3 (April 1, 1961): 360–84.
53. Rostow, *The Diffusion of Power*, 86, 111.
54. Ibid., 111, 87.
55. Rostow, "94. Rostow to Kennedy; February 28. 1961; Subj: Crucial Issues in Foreign Aid," 204.
56. US President's Committee to Study the US Military Assistance Program (1958–1959), "Draper Committee Final Report," 1, 11, 24–25, 36–37.
57. Amos et al., *U.S. Training of Foreign Military Personnel, Volume 2*, 103.
58. US President's Committee to Study the US Military Assistance Program (1958–1959), "Draper Committee Final Report," 39.
59. Ibid., 45.
60. Ibid., 43.
61. Ibid., 51.
62. Ibid., 43.
63. Ibid., 51.
64. "Old look" is Rostow's description in his memo to Kennedy. Rostow, "94. Rostow to Kennedy; February 28. 1961; Subj: Crucial Issues in Foreign Aid," 204.
65. Lerner, *The Passing of Traditional Society*; Gilman, *Mandarins of the Future*, 167, 139, 175.
66. Gilman, *Mandarins of the Future*, 158, 198.
67. Johnson, *The Role of the Military in Underdeveloped Countries*, 3.
68. Michel L. Martin, "Of Arms and the Man: A Short Intellectual History of Morris Janowitz's Contribution to the Sociology of the Military," in *The Military, Militarism, and the Polity: Essays in Honor of Morris Janowitz*, eds. Michel L. Martin and Ellen Stern McCrate (New York: Free Press, 1984), 1–31.
69. Henry Dodwell, *The Founder of Modern Egypt* (Cambridge: University Press, 1931). Dodwell uses the Muslim/Egyptian name "Muhammad Ali"; other scholars use the Turkish (and contemporary) "Mehmed"; Fahmy cites Abdel-Rahman al Rafi, "Mehmed Ali's Reign" (1951), Sabry's "L'Empire egyptien sous Mohamed-Ali en la question d'Orient" (1930), and Abdel-Rahman Zaki, "Military History of the Reign of Mehmed Ali the Great" (1950) as examples. "Take-off" is Fahmy's phrase, echoing Rostow's "Stages of Economic Growth." Khaled Fahmy, *All the Pasha's Men: Mehmed Ali, His Army, and the Making of Modern Egypt* (Cambridge: Cambridge University Press, 1997), 14, n. 37.
70. This anecdote appeared in Morroe Berger's CIS-sponsored 1960 monograph, *Military Elite and Social Change*. For his part, Gamal Nasser in 1955

considered Ali merely a corrupt foreigner who presided as Egyptians abandoned their "Eastern" national values while absorbing no "Western" ones. Berger, *Military Elite and Social Change*, 19; Gamal Abdel Nasser, "The Egyptian Revolution," *Foreign Affairs* 33, no. 2 (January 1, 1955): 199, 200.

71. Arnold Toynbee, "The Modernisation of the Middle East," *Journal of the Royal Institute of International Affairs* 8, no. 4 (July 1, 1929): 351.
72. Ibid.
73. Lewis V. Thomas and Richard Nelson Frye, *The United States and Turkey and Iran*, American Foreign Policy Library (Cambridge, MA: Harvard University Press, 1951); Halide Edib, *Turkey Faces West: A Turkish View of Recent Changes and Their Origin* (New Haven, CT: Yale University Press, 1930), 238.
74. Thomas and Frye, *The United States and Turkey and Iran.*
75. Quoted in Adas, *Dominance by Design*, 26.
76. Ekbladh, *The Great American Mission*, 18.
77. Ibid., 18–25; Stanley Karnow, *In Our Image: America's Empire in the Philippines* (New York: Random House, 1989).
78. Ekbladh, *The Great American Mission*, 20.
79. Alfred W. McCoy, *Closer Than Brothers: Manhood at the Philippine Military Academy* (New Haven, CT: Yale University Press, 1999), 18-20.
80. John Wands Sacca, "Like Strangers in a Foreign Land: Chinese Officers Prepared at American Military Colleges, 1904–37," *Journal of Military History* 70, no. 3 (July 1, 2006): 206. Ironically, Sacca argues, this liberal American education set them apart from their peers and superiors and, in the long run, hindered, rather than helped, their careers.
81. Frank L. Polk, *Invitation to Latin American Republics to Send Students to Military and Naval Academies* (Washington, DC: US Govt. Print. Off., 1919), 2.
82. To Authorize the President to Permit Citizens of the American Republics to Receive Instruction at Professional Education Institutions and Schools Maintained and Administered by the Government of the United States, Public Law 710, 75th Cong., 3d sess. (June 24, 1938), 1034.
83. Alfred Vagts, *A History of Militarism*, rev. ed. (London: Hollis and Carter, 1959), 14.
84. Ibid., 13.
85. Ibid., 490.
86. Martin, "Of Arms and the Man," 2.
87. Adas, *Machines as the Measure of Men.*
88. Reddy, "Modern," 160; Braham, *Key Concepts in Sociology*, 7.
89. Adas, *Dominance by Design*; Ekbladh, *The Great American Mission.*
90. Gilman, *Mandarins of the Future*, 32.

91. Shils, "The Military in the Political Development of the New States," 10–11.
92. See Bayly, *The Birth of the Modern World.*
93. Ibid., 10; Cooper warns against the mass confusion bound to ensue by calling modernity everything and everything modernity. See Cooper, *Colonialism in Question*, chap. 5.
94. Gilman, *Mandarins of the Future*, 32.
95. Cullather, "Modernization Theory." See, for example, Latham, *Modernization as Ideology*; Gilman, *Mandarins of the Future*; Tyrrell, *Reforming the World*; Nick Cullather, "The Foreign Policy of the Calorie," *American Historical Review* 112, no. 2 (April 1, 2007): 337–64.
96. Lerner, *The Passing of Traditional Society*, 78.
97. Latham, *Modernization as Ideology*, 8.
98. Millikan and Rostow, *A Proposal*, ix, 12.
99. Walt W. Rostow, *The Stages of Economic Growth: A Non-Communist Manifesto* (Cambridge: University Press, 1960).
100. James S. Coleman and Belmont Brice Jr., "The Role of the Military in Sub-Saharan Africa," in *The Role of the Military in Underdeveloped Countries*, ed. John J. Johnson (Princeton, NJ: Princeton University Press, 1962), 402.
101. Morris Janowitz, *The Military in the Political Development of New Nations: An Essay in Comparative Analysis* (Chicago: University of Chicago Press, 1964), 5.
102. Shils, "The Military in the Political Development of the New States," 89, 86.
103. Lerner and Robinson, "Swords and Ploughshares." See also, Dankwart A. Rustow, "The Army and the Founding of the Turkish Republic," *World Politics* 11, no. 4 (July 1, 1959): 513.
104. Pye, "Armies in the Process of Political Modernization," 69. An MIT professor of political science, Pye later taught courses in counterinsurgency for the State Department and advised USAID.
105. Pye, "Armies in the Process of Political Modernization." So also did: Lerner and Robinson, "Swords and Ploughshares." This article was reprinted (and discounted widely) in Henry Bienen, ed., *The Military and Modernization*, 1971. Shils, "The Military in the Political Development of the New States"; Janowitz, *The Military in the Political Development of New Nations.*
106. Latham, *Modernization as Ideology*, 7.
107. Gilman, *Mandarins of the Future*, 190.
108. Pye, "Armies in the Process of Political Modernization," 69. This 1962 volume printed the proceedings of the 1959 conference.
109. Ibid.
110. Lerner and Robinson, "Swords and Ploughshares," 34.

111. Janowitz, *The Military in the Political Development of New Nations*, vi.
112. Robert M. Price, "A Theoretical Approach to Military Rule in New States: Reference-Group Theory and the Ghanaian Case," *World Politics* 23, no. 3 (April 1, 1971): 399.
113. Shils, "The Military in the Political Development of the New States," 23.
114. Ibid.
115. Pye, "Armies in the Process of Political Modernization," 80.
116. Ibid., 74.
117. Ibid., 75.
118. Lerner and Robinson, "Swords and Ploughshares," 32.
119. Ibid. Emphasis in original.
120. Ibid., 34.
121. Ibid., 32.
122. Cullather, "Modernization Theory," 217; Lerner, *The Passing of Traditional Society*.
123. Lerner and Robinson, "Swords and Ploughshares," 13, 21.
124. Ibid., 21.
125. Ibid., 44.
126. Ibid.
127. Ibid., 41.
128. Rustow, "The Army and the Founding of the Turkish Republic," 513.
129. Amin Banani, *The Modernization of Iran* (Stanford, CA: Stanford University Press, 1961).
130. Berger, *Military Elite and Social Change*, 6.
131. Ibid.
132. Ibid., 3. Emphasis is mine.
133. Nick Cullather, "'Fuel for the Good Dragon': The United States and Industrial Policy in Taiwan, 1950–1965," *Diplomatic History* 20, no. 1 (1996): 11.
134. Cullather, "Modernization Theory," 213.
135. Pye, "Armies in the Process of Political Modernization," 82.
136. Coleman and Brice Jr., "The Role of the Military in Sub-Saharan Africa," 402.
137. Lerner and Robinson, "Swords and Ploughshares," 32.
138. Janowitz, *The Military in the Political Development of New Nations*, 63. Since then, many others have refuted the claim, including Altnay (for Turkey) and Fahmy (for Egypt). Ayşe Gül Altınay, *The Myth of the Military Nation: Militarism, Gender, and Education in Turkey* (New York: Palgrave Macmillan, 2004); Fahmy, *All the Pasha's Men*.

139. Pye, "Armies in the Process of Political Modernization," 77.
140. Ibid., 87.
141. Ibid.
142. Rostow, *The Process of Economic Growth*, 26.
143. Pye, "Armies in the Process of Political Modernization," 77.
144. Gutteridge, *Military Institutions and Power in the New States*, 44.
145. Janowitz, *The Military in the Political Development of New Nations*, 60.
146. Lerner and Robinson, "Swords and Ploughshares," 34; Coleman and Brice Jr., "The Role of the Military in Sub-Saharan Africa," 401; Pye, "Armies in the Process of Political Modernization," 80. See also David Killingray, "Imagined Martial Communities: Recruiting for the Military and Police in Colonial Ghana, 1860–1960," in *Ethnicity in Ghana: The Limits of Invention*, eds. Carola Lentz and Paul Nugent (New York: St. Martin's Press, 2000), 119; Timothy Parsons, *The African Rank-and-File: Social Implications of Colonial Military Service in the King's African Rifles, 1902–1964*, Social History of Africa Series (Portsmouth, NH: Heinemann, 1999); Myron J. Echenberg, *Colonial Conscripts: The Tirailleurs Sénégalais in French West Africa, 1857–1960*, Social History of Africa Series (Portsmouth, NH: Heinemann, 1991).
147. S. Kojo Addae, *A Short History of Ghana Armed Forces* (Accra: Ministry of Defence of Ghana Armed Forces, 2005), 170; Akwasi A. Afrifa, *The Ghana Coup, 24th February 1966* (London: Cass, 1966), 93.
148. Addae, *A Short History of Ghana Armed Forces*, xii. See also, Killingray, "Imagined Martial Communities," 129; David Killingray, "Soldiers, Ex-Servicemen, and Politics in the Gold Coast, 1939–50," *Journal of Modern African Studies* 21, no. 03 (1983): 523–34.
149. Lerner and Robinson, "Swords and Ploughshares," 34. Emphasis in original.
150. Ibid., 38.
151. Janowitz, *The Military in the Political Development of New Nations*, 97.
152. Lerner and Robinson, "Swords and Ploughshares," 39.
153. Ibid.
154. Mazower, *Governing the World*, 291.
155. Quoted in Latham, *Modernization as Ideology*, 58. On "court historian," see Sam Tanenhaus, "History, Written in the Present Tense," *New York Times*, March 4, 2007, Week in Review.
156. Rostow, "94. Rostow to Kennedy; February 28. 1961; Subj: Crucial Issues in Foreign Aid," 207.
157. Latham, *Modernization as Ideology*, 55, 7.
158. Rostow, "94. Rostow to Kennedy; February 28. 1961; Subj: Crucial Issues in Foreign Aid," 205.
159. Windle and Vallance, "Optimizing Military Assistance Training," 96, 92.

160. Harlan Cleveland, *The Overseas Americans*, 1st ed., The Carnegie Series in American Education (New York: McGraw-Hill, 1960), 79.
161. John F. Kennedy, "Final Annual Report on the Operations of the Mutual Security Program," H.R. Doc. No. 87-432 at 2 (June 12, 1962); Kennedy, "Special Message to the Congress on Foreign Aid—March 22, 1961," 205.
162. Dean Rusk, "95. Memorandum from Secretary of State Rusk to President Kennedy; March 10, 1961," in *FRUS 1961–1963*, vol. ix, *Foreign Economic Policy*, eds. Evans Gerakas et al. (Washington, DC: US Govt. Print. Off., 1995), 210.
163. Ibid.
164. Kennedy, "Special Message to the Congress on Foreign Aid—March 22, 1961."
165. Ibid., 208.
166. *Hearings Before the Senate Committee On Foreign Relations on the Foreign Assistance Act of 1961*, 28.
167. Ibid.
168. Chester C. Bowles, "30. Memorandum from the Under Secretary of State (Bowles) to Secretary of State Rusk; Subject: The Need to Improve the Administration of Foreign Policy; June 14, 1961," in *FRUS 1961–1963*, vol. xxv, *Organization of Foreign Policy; Information Policy; United Nations; Scientific Matters*, eds. Paul Claussen, Evan M. Duncan, and Jeffrey A. Soukup (Washington, DC: US Govt. Print. Off., 2001), 56.
169. Ibid.
170. See William O. Webb, "84. Report Prepared by the Policy Planning Council; Undated (c. January 19, 1961); Subject: US Military Aid Policy toward Non-NATO Countries," in *Foreign Economic Policy*, vol. ix, Foreign Relations of the United States, 1961–1963 (Washington DC: US Govt. Print. Off., 1995), 191.
171. Ibid., 189.
172. Roswell Gilpatrick, "Memorandum for the Service Secretaries from the Deputy Secretary of Defense, September 13, 1963; Subject: An Informational Program for Foreign Military Trainees and Visitors in the United States" (Memo, Washington, DC, September 13, 1963), FOIA request, in author's possession.
173. Ibid., 2.
174. Ibid.
175. US Department of Defense, "Department of Defense Directive 5410.17 'An Informational Program for Foreign Military Trainees and Visitors in the United States,'" January 15, 1965, Combined Arms Research Library; US Department of Defense, "Department of Defense Directive 5410.17

'An Informational Program for Foreign Military Trainees and Visitors in the United States,'" March 1, 1985, accessed March 18, 2012, *www.dtic.mil/dtic/tr/fulltext/u2/a272252.pdf*; US Department of Defense, "Department of Defense Instruction 5410.17 United States Field Studies Program (FSP) for International Military and Civilian Students and Military-Sponsored Visitors," September 15, 2006, accessed March 18, 2012, *www.dtic.mil/whs/directives/corres/pdf/541017p.pdf*.

176. US Department of Defense, "DoDD 5410.17, 15 January 1965," 3.
177. Gilpatrick, "September 13, 1963 Memo: Informational Program for Foreign Military Trainees."
178. Gilman, *Mandarins of the Future*, 20.
179. "Sample Activities" and "Visiting Lecturers" slides in William H. Eggering, "Class Director's Mission Statement" (presentation, September 10, 1991), box 1, folder 6, CARL International Officer Collection.
180. "Commandant's Goals" slide in ibid.; for more on Leavenworth and Kansas City sponsor programs, see John V. Clune, "The Abongo Abroad: Military Internationalism, Travel, Training, and Peace in Ghana and the United States, 1960–1992" (PhD Dissertation, University of Kansas, 2014), chap. 4.
181. Windle and Vallance, "Optimizing Military Assistance Training," 91; William Philip Pipkin, "Allied Personnel at USACGSC," *Reconnaissance*, 1980, box 1, folder 8, CARL International Officer Collection.
182. "PTP Program and Goals Emphasized at First Installation Dinner," *Contact: Greater Kansas City People-to-People Council Newsletter*, April 1963, GKCPTP; John Reichley, *International Officers: A Century of Participation at the United States Army Command and General Staff College* (Leavenworth, KS: US Army Command and General Staff College, 1994), 24; Col. (Ret.) William P. Pipkin, "Programs for Allied Officers," c 1992, box 1, folder 3, CARL International Officer Collection.
183. Dan Millman, "Inside PTP," *Contact: Greater Kansas City People-to-People Council Newsletter*, October 1969, binder: Contact 1963–1970, GKCPTP.
184. Latham, *Modernization as Ideology*, chap. 3; General Accounting Office, *School of the Americas: US Military Training for Latin American Countries; Report to the Ranking Minority Member, Committee on National Security, House of Representatives* (Washington, DC: GAO, National Security and International Affairs Division, August 22, 1996), 20. The US Army Caribbean School–Spanish Instruction existed there since 1946, Kennedy renamed it the School of the Americas in 1963. It moved to Fort Benning, Georgia, in 1984; The US Air Force ran a smaller Inter-American Air Forces Academy and the US Navy ran the Navy Small Craft Instruction and Technical Team from the Canal Zone as well. Amos et al., *U.S. Training of Foreign Military Personnel, Volume 2*, 62.

185. Fiscal years 1962–1970 for Africa: $19.4 million; for Latin America: $91.3 million. Total number of students from Africa FY 1962–1970: 3,348 ($11.6 million and 1,900 of which were from Ethiopia, where the United States had a mobile training team); from Latin America: 43,831. Amos et al., *U.S. Training of Foreign Military Personnel, Volume 2*, 99–110.
186. Ibid., 63.
187. General Accounting Office, *School of the Americas*, 9.
188. Lawrence Weschler, "A Reporter at Large: The Great Exception I—Liberty," *New Yorker*, April 3, 1989, 48.
189. Jack Nelson-Pallmeyer, *School of Assassins: Guns, Greed, and Globalization*, rev. ed. (Maryknoll, NY: Orbis Books, 2001), 19.
190. Lesley Gill, *The School of the Americas: Military Training and Political Violence in the Americas*, American Encounters/Global Interactions Series (Durham, NC: Duke University Press, 2004), 28.
191. Ibid., 33.
192. Westad, *The Global Cold War*, 5.
193. See Hoffman, *All You Need Is Love*.
194. In Africa, there were at least eighteen coups or attempted coups between 1962 and 1967; A. B. Assensoh and Yvette Alex-Assensoh, *African Military History and Politics: Ideological Coups and Incursions, 1900–Present* (New York: Palgrave Macmillan, 2002), 7.
195. Henry Bienen, ed., *The Military and Modernization* (Chicago: Aldine Atherton, 1971), 9. See also, John P. Lovell and C. I. Eugene Kim, "The Military and Political Change in Asia," *Pacific Affairs* 40, no. 1/2 (April 1, 1967): 113–23; Price, "A Theoretical Approach to Military Rule in New States"; Robert M. Price, "Military Officers and Political Leadership: The Ghanaian Case," *Comparative Politics* 3, no. 3 (April 1, 1971): 361–79.
196. On Turkey, see Claude E. Welch Jr., "The African Military and Political Development," in *The Military and Modernization*, ed. Henry Bienen (Chicago: Aldine Atherton, 1971), 224. On military rulers' persistence, see Bienen, *The Military and Modernization*. On "theory-research gap," see Price, "A Theoretical Approach to Military Rule in New States." On administration skills, see Welch, "The African Military and Political Development," 228; Price, "Military Officers and Political Leadership"; Aristide R. Zolberg, "The Structure of Political Conflict in the New States of Tropical Africa," *American Political Science Review* 62, no. 1 (March 1, 1968): 75.
197. The University of Chicago's Aristide Zolberg even suggested that the coup had become the normal, if not preferred, method of regime change in African politics—the functional equivalent of elections. Zolberg, "The Structure of Political Conflict in the New States of Tropical Africa," 77–78.

198. Huntington, "Political Development and Political Decay," 386.
199. Ibid., 423.
200. Ibid., 420.
201. Gilman, *Mandarins of the Future*, 229.
202. Ibid., 205.
203. Chomsky, *American Power and the New Mandarins*, 332.
204. Andre Gunder Frank, "The Development of Underdevelopment," *Monthly Review* 18, no. 4 (1966): 17–31; Joyce Kolko and Gabriel Kolko, *The Limits of Power: The World and United States Foreign Policy, 1945–1954* (New York: Harper & Row, 1972), 2; Williams, *The Tragedy of American Diplomacy*, 14; Teresa Hayter, *Aid as Imperialism.* (Baltimore: Penguin, 1971).
205. Péter Tamás Bauer, *Dissent on Development; Studies and Debates in Development Economics* (Cambridge, MA: Harvard University Press, 1972).
206. Vietnam-related grant military aid usually does not appear in annual foreign aid reports because it was provided from a separate military assistance service fund account, which was outside the regular Military Assistance Program account. Grimmett, *Foreign Assistance and Related Programs Appropriations for 1986*, 430; Ruttan, *United States Development Assistance Policy: The Domestic Politics of Foreign Economic Aid*, 105.
207. H. Comm. on Foreign Affairs, Foreign Assistance Act of 1969; Report of the Committee on Foreign Affairs Together with Minority, Supplemental and Additional Views on H.R. 14580 to Promote the Foreign Policy, Security, and General Welfare of the United States by Assisting Peoples of the World, Etc., H.R. Rep. No. 91-611, at 10 (November 6, 1969).
208. Chomsky, *American Power and the New Mandarins*; Gilman, *Mandarins of the Future*; Simpson, *Economists with Guns*. For Rostow on Vietnam in 1974, see *Hearts and Minds*, dir. by Peter Davis, BBS Productions, 1974 (Criterion Collection, 2002), DVD, 1.1:32; W. W. Rostow, "Vietnam and Asia," *Diplomatic History* 20, no. 3 (1996): 268. See also Nick Cullather, "Research Note: Development? It's History," *Diplomatic History* 24, no. 4 (October 2000): 641–53.
209. Donald M. Fraser, "New Directions in Foreign Aid," *World Affairs* 129, no. 4 (January 1, 1967): 245, 250.
210. Ibid., 249–250.
211. Robert S McNamara, "Draft Memorandum for the President on the Military Assistance Program, FY 1969–1973; December 18, 1967" (Washington, DC, December 18, 1967), 3–7, RG 59 CFPF 1967–1969, box 1684, folder DEF 19 Ghana, NACP.
212. Ibid., 8.
213. Ibid., 9.
214. Ibid., 11.
215. Ibid., 12.

216. Ibid.
217. Ibid.
218. Total MAP numbers from United States Defense Security Assistance Agency (DSAA), *Congressional Presentation, Security Assistance Programs*, FY 1981 (1980), 6. MAP did not make up the entirety of US security assistance. Many different programs fell under this broader umbrella, including vast sums of military and economic aid for Egypt and Israel. Also, specific labels changed over the years. For consistency, I use the term "MAP" as a relatively stable benchmark for a specific set of military assistance programs that exclude other forms of economic aid.
219. Ibid.
220. Gilman, *Mandarins of the Future*, 23.
221. Hoffman, *All You Need Is Love*, 9.
222. Latham, *Modernization as Ideology*, 59.
223. Cullather, *The Hungry World*, chap. 4, 5, 6, 8.
224. David Milne, *America's Rasputin: Walt Rostow and the Vietnam War* (New York: Hill and Wang, 2008); Cullather, *The Hungry World*, 229.

## CHAPTER 2

1. Charles de Graft-Dickson, "Speech by the Minister Responsible for Defense on the Occasion of the Presentation of Books to the Ghana Military Academy by the Canadian Government" (Ghana Military Academy, Teshie, August 17, 1961), RG 14/4/506 Ghana Military Academy, doc. 188, PRAAD. Dickson appears as Dixon in other sources.
2. Ibid.
3. President Kwame Nkrumah flamboyantly christened himself Ghana's *Osagyefo*—usually translated "savior" or "messiah," although de Graft-Dickson says "the Osagyefo, the President" here. Most documents in this RG use the term when referring to Nkrumah in the third person. Ibid.
4. Harry R. Coates, "Letter from Lt. Col. Harry R. Coates (US Army) to N. F. Ribeiro-Ayeh (Ministry of Defense); August 18, 1961; Subject: Books and Pamphlets" (US Embassy, Accra, August 18, 1961), RG 14/4/506 Ghana Military Academy, doc. 189, PRAAD.
5. Ibid.
6. "Michel Dies on Day of Triumph," *Daily Graphic*, September 3, 1961, 1, CAMP Reel 12, July–Sept. 1961, Center for Research Libraries.
7. Brigadier equates to the American rank of brigadier general. Aboagye, *The Ghana Army*, 268, 276; Addae, *A Short History of Ghana Armed Forces*, 171, 177, 388; Francis Ameyibor, "Ghana Remembers Brigadier Joseph Michel's 50 Years Anniversary," Ghanaian News, *GhanaWeb* (September 7, 2011), *www.ghanaweb.com/GhanaHomePage/features/artikel.php?ID=218069*. Oddly,

all three of these sources list a different month for the plane crash, July, August, and September 1961.

8. Henry Kwami Anyidoho, *My Journey . . . Every Step* (Accra, Ghana: Sub-Saharan Publishers, 2010), 24; "Ghana Paper Charges Britain Plotted Hammarskjold's Death," *New York Times*, September 21, 1961, ProQuest Historical Newspapers; "A Congo Martyr," *West Africa*, no. 2312 (September 23, 1961): 1057.
9. Russell, "News Release by United States Embassy, Accra; September 11, 1961; Subject: Remarks of the American Ambassador Mr. Francis K. Russell at the Presentation of Books to the Ghana Military Academy."
10. Ibid.
11. Maj. Gen. S. J. A. Otu, "Loose Minute; May 1, 1962; Subject: Gift of Russian Books for GMA Library" (Burma Camp, Accra, May 1, 1962), RG 14/4/506 Ghana Military Academy, doc. 234, PRAAD; Maj. Gen. S. J. A. Otu, "P.S.; May 3, 1962; Subject: Reference Minute 4 on Page 234" (Burma Camp, Accra, May 3, 1962), RG 14/4/506 Ghana Military Academy, doc. 235, PRAAD; "Message from the Principal Secretary, Ministry of Foreign Affairs (Accra) to the Principal Secretary, Ministry of Defense (Accra); August 14, 1962; Subject: Gift of Books to the Ghana Military Academy By the Army of the Soviet Union" (Burma Camp, Accra, August 14, 1962), RG 14/4/506 Ghana Military Academy, doc. 237 (approximately—damaged), PRAAD; A. A. Adusei, "Message from the Principal Secretary, Ministry of Defense (Accra) to the Principal Secretary, Ministry of Foreign Affairs (Accra); October 4, 1962; Subject: Gift of Books to the Ghana Military Academy By the Army of the Soviet Union" (Burma Camp, Accra, October 4, 1962), RG 14/4/506 Ghana Military Academy, doc. 274, PRAAD; "Message from the Principal Secretary, Ministry of Foreign Affairs (Accra) to the Principal Secretary, Ministry of Defense (Accra); November 1, 1962; Subject: Gift of Books to the Ghana Military Academy By the Army of the Soviet Union" (Burma Camp, Accra, November 1, 1962), RG 14/4/506 Ghana Military Academy, doc. 284, PRAAD.
12. "Ministry of Foreign Affairs to Ministry of Defense, November 1, 1962; Subj: Gift of Books to GMA by USSR."
13. Kwame Nkrumah, *The Autobiography of Kwame Nkrumah* (London: Panaf, 1973), 136; "Ghana Asks 'Bold' Western Aid," *New York Times,* July 6, 1960, ProQuest Historical Newspapers; Apter, *The Gold Coast in Transition* re-published in 1963 as "Ghana in Transition," with new editions in 1968 and 1972. Baynham, *The Military and Politics in Nkrumah's Ghana.*
14. Baynham, *The Military and Politics in Nkrumah's Ghana.*
15. H. T. Alexander, *African Tightrope: My Two Years as Nkrumah's Chief of Staff* (London: Pall Mall, 1965).

16. Baynham, *The Military and Politics in Nkrumah's Ghana*; Addae, *A Short History of Ghana Armed Forces*, 134, 144; Aboagye, *The Ghana Army*, 98.
17. Addae, *A Short History of Ghana Armed Forces*, 170–186.
18. Ama Biney, *The Political and Social Thought of Kwame Nkrumah* (New York: Palgrave Macmillan, 2011), 143; Apter, *Ghana in Transition*, 385; Alexander, *African Tightrope*, chap. 4.
19. Embassy of Ghana, "Ghana to be Self-Sufficient by 1977—Acheampong," *Ghana News*, October 15, 1974; Pte. David Gamor, "OFY . . . Big Plan for Military," *Armed Forces News*, August 31, 1974, Ghana Armed Forces Director of Public Relations (Hereafter: GAF DPR); Eben Quarcoo, "OFY [Operation Feed Yourself] Grips Forces," *Armed Forces News*, February 15, 1975; Embassy of Ghana, "Apartheid in South Africa: Dr. Busia Calls for Additional Methods," *Ghana News*, February 1971.
20. Research and Development Department, ed., *Ghana Armed Forces Staff College: An Historical Perspective* (Accra, Ghana: Jamieson's Cambridge Faxbooks, 1996), 1.
21. National Security Council, "245. Ghana: Assessment Since Volta; June 13, 1962," in *FRUS 1961–1963*, vol. xxi, *Africa*, ed. Nina Davis Howland (Washington, DC: US Govt. Print. Off., 1995), 375–77; NSC Interdepartmental Group for Africa, "Secret Ghana Policy Planning Paper by NSC Interdepartmental Group for Africa, May 1971" (Report, Washington, DC, May 1971), RG 59 SNF 1970–1973, box 2325, folder POL 1 Ghana-US 10/15/70, NACP; DSAA, *Congressional Presentation, Security Assistance Programs*, FY 1981, 277.
22. Valerie Plave Bennett, "The Motivation for Military Intervention: The Case of Ghana," *Western Political Quarterly* 26, no. 4 (December 1, 1973): 659–74; Dennis Austin and Robin Luckham, eds., *Politicians and Soldiers in Ghana, 1966–1972*, Studies in Commonwealth Politics and History 3 (London: Cass, 1975).
23. Bennett, "The Motivation for Military Intervention"; Naomi Chazan, *An Anatomy of Ghanaian Politics: Managing Political Recession, 1969–1982* (Boulder, CO: Westview Press, 1983).
24. Benon Tumukunde, "Inside a 'Colonised Mind,'" *The Flame: The Magazine of the GAFCSC*, August 1989, 70–72, GAFCSC Library. See Ngugi wa Thiongo, *Decolonising the Mind: The Politics of Language in African Literature* (London: J. Currey, 1986).
25. Tumukunde, "Inside a 'Colonised Mind,'" 72.
26. Westad, *The Global Cold War*, 5; Elizabeth Cobbs Hoffman, "Diplomatic History and the Meaning of Life: Toward a Global American History," *Diplomatic History* 21, no. 4 (1997): 79; Painter, "Explaining U.S. Relations with the Third World," 533.

27. Schmidt, "Nkrumah Confers at White House; Backs U.N. Effort"; Rostow, *The Diffusion of Power*, 201–203; Kwame Nkrumah, *I Speak of Freedom: A Statement of African Ideology* (London: Heinemann, 1961); Kwame Nkrumah, *Africa Must Unite*, new edition (New York: International Publishers, 1970); Bundy, "239. Bundy to Kennedy; December 1, 1961; Subj: Volta Project"; Kennedy, "242. Letter from President Kennedy to President Nkrumah; December 14, 1961."
28. Painter, "Explaining U.S. Relations with the Third World," 533–535.
29. Rostow, *The Diffusion of Power*, 203.
30. Muehlenbeck, *Betting on the Africans*, chap. 4.
31. Mazower, *Governing the World*, 296; Robert O. Collins and James M. Burns, *A History of Sub-Saharan Africa*, 2nd ed. (New York: Cambridge University Press, 2013), 363. The verdict is still out on the long-term environmental consequences. See P. Gyau-Boakye, "Environmental Impacts of the Akosombo Dam and Effects of Climate Change on the Lake Levels," *Environment, Development and Sustainability* 3, no. 1 (2001): 17.
32. Walt W. Rostow, "235. Memorandum from the President's Deputy Special Assistant for National Security Affairs (Rostow) to President Kennedy; Subject: The Volta Project; October 2, 1961," in *FRUS 1961–1963*, vol. xxi, *Africa*, ed. Nina Davis Howland (Washington, DC: US Govt. Print. Off., 1995), 359; Penfield, "223. Memorandum from the Deputy Assistant Secretary of State for African Affairs (Penfield) to the Under Secretary of State (Bowles); Feb. 17, 1961; Subject: The Volta River and Related Projects," in *FRUS 1961–1963*, vol. xxi, *Africa*, ed. Nina Davis Howland (Washington, DC: US Govt. Print. Off., 1995), 341–342; Dean Rusk, "236. Telegram From the Department of State to the Embassy in Chad; October 10, 1961," in *FRUS 1961–1963*, vol. xxi, *Africa* ed. Nina Davis Howland (Washington, DC: US Govt. Print. Off., 1995), 360–61; "Ghana Dam Explosive Threat in West Africa: Parallel to Egypt's Aswan Dam Seen as U.S. Reviews Plans to Finance Project," *Los Angeles Times*, October 26, 1961, ProQuest Historical Newspapers. Richard D. Mahoney, *JFK: Ordeal in Africa* (New York: Oxford University Press, 1983), 173.
33. "232. Kennedy to Ball; September 21, 1961," 355.
34. Rusk, "236. Telegram from the Department of State to the Embassy in Chad; October 10, 1961"; Bundy, "239. Bundy to Kennedy; December 1, 1961; Subj: Volta Project"; Bundy, "240. Notes for the Record; December 5, 1961; Subject: NSC Meeting on Volta Dam," 369; Kennedy, "242. Letter from President Kennedy to President Nkrumah; December 14, 1961," 372.
35. Cobbs, "Decolonization, the Cold War"; Muehlenbeck, *Betting on the Africans*; Bundy, "239. Bundy to Kennedy; December 1, 1961; Subj: Volta Project"; Meriwether also argues that Kennedy substituted aggressive African

diplomacy for legitimate leadership in domestic civil rights issues in "'Worth a Lot of Negro Votes.'"

36. Rostow, *The Diffusion of Power*, 200.
37. Webb, "84. Report Prepared by the Policy Planning Council, c. January 19, 1961," 191; Office of the Director of Military Assistance, OASD (IS), "Military Assistance Plan: Ghana (Secret)," July 15, 1963, A1, RG 330, accession no. UD/WW 241, box 14, folder "Ghana Five Year Military Assistance Plans 1965–1969," NACP.
38. Office of the Director of Military Assistance, OASD (IS), "Military Assistance Plan: Ghana (Secret)," A1.
39. Ibid., D2–D3.
40. Kolko, *Confronting the Third World*, 111. I argue that the United States did not defer to the colonial power so much as incorporate the colonial power in its own strategy when the results seemed promising. See also, Painter, "Explaining U.S. Relations with the Third World."
41. James M. Gavin, "Confidential Telegram from James M. Gavin, American Ambassador to France to Secretary of State; Number 4551, March 28 1962," March 28, 1962, RG 59 CDF 1960–1963, box 1482; folder 711.55341/1-2260, doc. 711.55370/3-2862, NACP.
42. Ibid.
43. Painter, "Explaining U.S. Relations with the Third World," 533.
44. Office of the Director of Military Assistance, OASD (IS), "Military Assistance Plan: Ghana (Secret)," D2.
45. Ibid., C1.
46. Ibid., D1.
47. In 1962 US-dollar equivalents. Ibid., J1.
48. Ibid., D2.
49. Ibid., D3.
50. Ibid.
51. Office of the Director of Military Assistance, OASD (IS), "Military Assistance Plan Draft: Ghana (Secret)," April 11, 1963, 7, RG 330, accession no. UD/WW 241, box 14, folder "Ghana Five Year Military Assistance Plans 1965–1969," NACP.
52. Office of the Director of Military Assistance, OASD (IS), "Military Assistance Plan: Ghana (Secret)," C1.
53. Ibid., C1.
54. Ibid.
55. Office of the Director of Military Assistance, OASD (IS), "Military Assistance Plan Draft: Ghana (Secret)," sec. 3, 7. Extra comma appears in original.
56. Office of the Director of Military Assistance, OASD (IS), "Military Assistance Plan: Ghana (Secret)," D3.

57. Ibid., G1.
58. Office of the Director of Military Assistance, OASD (IS), "Military Assistance Plan Draft: Ghana (Secret)," sec. 3, 7.
59. Office of the Director of Military Assistance, OASD (IS), "Military Assistance Plan: Ghana (Secret)," G1, L2.
60. Office of the Director of Military Assistance, OASD (IS), "Military Assistance Plan Draft: Ghana (Secret)," sec. 3.
61. Keith Hopkins, "Civil-Military Relations in Developing Countries," *British Journal of Sociology* 17, no. 2 (June 1, 1966): 165–82; Gutteridge, *Military Institutions and Power in the New States*; William F. Gutteridge, "The Political Role of African Armed Forces: The Impact of Foreign Military Assistance," *African Affairs* 66, no. 263 (April 1, 1967): 93–103; Raymond A. Moore, "The Peaceful Uses of Military Forces in Underdeveloped Areas: A Review Essay," *Journal of Developing Areas* 4, no. 1 (October 1, 1969): 112–19; Claude E. Welch Jr., ed., *Soldier and State in Africa: A Comparative Analysis of Military Intervention and Political Change* (Evanston, IL: Northwestern University Press, 1970); Ernest W. Lefever, *Spear and Scepter: Army, Police, and Politics in Tropical Africa* (Washington, DC: Brookings Institution, 1970).
62. "Nkrumah as charismatic leader" was not a new trope in 1966. See Lionel Tiger, "Nkrumah's Ghana and the Theory of Charisma," *Bulletin of African Studies in Canada / Bulletin Des Études Africaines Au Canada* 2, no. 1 (November 1, 1964): 2–10. See also David E. Apter, "Nkrumah, Charisma, and the Coup," *Daedalus* 97, no. 3 (July 1, 1968): 757–92; Jon Kraus, "Arms and Politics in Ghana," in *Soldier and State in Africa: A Comparative Analysis of Military Intervention and Political Change*, ed. Claude E. Welch (Evanston, IL: Northwestern University Press, 1970), 154–223; Price, "A Theoretical Approach to Military Rule in New States"; Price, "Military Officers and Political Leadership"; Selwyn Ryan, "The Theory and Practice of African One Partyism: The CPP Re-Examined," *Canadian Journal of African Studies / Revue Canadienne Des Études Africaines* 4, no. 2 (April 1, 1970): 145–72.
63. Kwame Nkrumah, *Dark Days in Ghana* (London: Panaf, 1968); Nkrumah, *Africa Must Unite*; Kwame Nkrumah, *Class Struggle in Africa*, 1st ed. (New York: International Publishers, 1970); Nkrumah, *Revolutionary Path*; Afrifa, *The Ghana Coup, 24th February 1966*; A. K. Ocran, *Politics of the Sword: A Personal Memoir on Military Involvement in Ghana and of Problems of Military Government* (London: Rex Collings, 1977).
64. Gutteridge, "The Political Role of African Armed Forces," 96.
65. Numbers of officers from Aboagye, *The Ghana Army*, 119; "Charter of the Ghana Military Academy; August 26, 1959"; Lt. Col. A. M. Gilmour,

"Memorandum from Lt Col A. M. Gilmour, Chief of the General Staff, Ghana Armed Forces to J. S. Annan, Permanent Secretary, Ministry of Foreign Affairs; Subject: Entry To GMA [Ghana Military Academy]" (Ministry of Defense, Giffard Camp, Accra, January 30, 1960), RG 14/4/504 Ghana Military Academy, doc. 62, PRAAD; Addae, *A Short History of Ghana Armed Forces*, 125; Huntington, *The Soldier and the State*, chap. 1; Alexander, *African Tightrope*, chap. 2.

66. Addae, *A Short History of Ghana Armed Forces*, 127; Aboagye, *The Ghana Army*, 119.
67. Nkrumah, *Africa Must Unite*; Alexander, *African Tightrope*, 33–49.
68. Aboagye, *The Ghana Army*, 192. Regular Officers' Special Training School (ROSTS) prepared officer candidates for training at European military schools. See also Baynham, *The Military and Politics in Nkrumah's Ghana*, 32–64.
69. That one was future chief of defense staff Okai, who graduated from Sandhurst in 1955. Addae, *A Short History of Ghana Armed Forces*, 165–169; Aboagye, *The Ghana Army*, 112; Lawrence Adjei, interview with author, audio recording, July 20, 2012, author's possession.
70. Addae, *A Short History of Ghana Armed Forces*, 125.
71. Gilmour, "Gilmour to Annan; Sept. 1, 1960; Subj: Entry to GMA"; Aboagye, *The Ghana Army*, 113.
72. Gilmour, "Gilmour to Annan; Sept. 1, 1960; Subj: Entry to GMA." In the archive, Gilmour's memorandum is the cover sheet for the entire plan.
73. The GMA projected to lose between 20 and 30 percent of each class due to attrition. "Charter of the Ghana Military Academy; August 26, 1959."
74. For instance, Anyidoho refers to his comrades in "regular intake VIII"; Frimpong identifies the first female cadets by intake number; Oduro-Kwarteng identifies those in his intake who rose to prominent positions, as well as those in the intakes just prior to and after his own. Anyidoho, *My Journey . . . Every Step*, 34; Brig. Gen. Daniel Kwadjo Frimpong, *Leadership and the Challenges of Command: The Ghana Military Academy Experience* (Accra, Ghana: Afram Publications, 2003), 49; Steve Oduro-Kwarteng, *The Memoirs of a Colonel Retired* (Accra, Ghana: self-published, 2009), 14.
75. "Charter of the Ghana Military Academy; August 26, 1959."
76. Gilmour, "Gilmour to Annan; Sept. 1, 1960; Subj: Entry to GMA."
77. Ibid.
78. Ibid.
79. Ibid.

80. Addae, *History of Ghana Armed Forces*, 3:91.
81. Ignatius Achel, interview with author, audio recording, July 19, 2012, 3, author's possession.
82. Nkrumah, *I Speak of Freedom*, 128. See also, Ryan M. Irwin, *Gordian Knot: Apartheid and the Unmaking of the Liberal World Order*, Oxford Studies in International History (Oxford: Oxford University Press, 2012), 32–39.
83. Nkrumah, *I Speak of Freedom*, 211–222.
84. Ibid., 219.
85. Joseph S. Annan, "Memorandum from Joseph S. Annan, Permanent Secretary, Ministry of Defense, to the Permanent Secretary, Ministry of Education and Information, September 24, 1959; Subject: Graduate Teachers for the Ghana Military Academy" (Ministry of Defense, Giffard Camp, Accra, September 24, 1959), RG 14/4/504 Ghana Military Academy, doc. 10, PRAAD.
86. Ibid.
87. Christopher R. Kilford, *The Other Cold War: Canada's Military Assistance to the Developing World, 1945–1975* (Kingston, Ontario: Canadian Defence Academy Press, 2010), 103; Aboagye, *The Ghana Army*, 193.
88. Killingray, "Imagined Martial Communities," 123; See also Parsons, *The African Rank-and-File*, chap. 3; Echenberg, *Colonial Conscripts*, 14.
89. Killingray, "Imagined Martial Communities," 125.
90. Ibid., 129.
91. Parsons, *The African Rank-and-File*, chap. 5; Echenberg, *Colonial Conscripts*; Killingray, "Imagined Martial Communities," 127; Gregory Mann, *Native Sons: West African Veterans and France in the Twentieth Century* (Durham, NC: Duke University Press, 2006).
92. Carolyn A. Brown, "A 'Man' in the Village Is a 'Boy' in the Workplace: Colonial Racism, Worker Militance, and Igbo Notions of Masculinity in the Nigerian Coal Industry, 1930–1945," in *Men and Masculinities in Modern Africa*, eds. Lisa A. Lindsay and Stephan Miescher (Portsmouth, NH: Heinemann, 2003), 156–74; Frederick Cooper, "Industrial Man Goes to Africa," in *Men and Masculinities in Modern Africa*, eds. Lisa A. Lindsay and Stephan Miescher (Portsmouth, NH: Heinemann, 2003), 128–38; Even Mehmet's pseudomodern Egyptian army attempted such bodily control, with poor results. Fahmy, *All the Pasha's Men*, 221–235.
93. Anderson, *Imagined Communities*, 116–121.
94. "Charter of the Ghana Military Academy; August 26, 1959."
95. Ibid.; By 1961, fourteen Ghanaian women served as officers (all nurses). Ghana Military Academy did not accept women until 1964. Aboagye, *The Ghana Army*, 113.

96. "Charter of the Ghana Military Academy; August 26, 1959." Transcription closely matched to the original.
97. On language and nationalism, see Anderson, *Imagined Communities*, 133; Chinua Achebe, "English and the African Writer," *Transition*, no. 18 (January 1, 1965): 27–30. On twentieth-century internationalism and transnationalism, see Daniel T. Rodgers, *Atlantic Crossings: Social Politics in a Progressive Age* (Cambridge, MA: Belknap Press of Harvard University Press, 1998); Daniel Gorman, *The Emergence of International Society in the 1920s* (Cambridge: Cambridge University Press, 2012); Tyrrell, *Reforming the World*; Mazower, *Governing the World*; Ekbladh, *The Great American Mission*; Erez Manela, *The Wilsonian Moment: Self-Determination and the International Origins of Anticolonial Nationalism* (Oxford: Oxford University Press, 2009); Andrew Zimmerman, *Alabama in Africa: Booker T. Washington, the German Empire, and the Globalization of the New South* (Princeton, NJ: Princeton University Press, 2010); Christopher Endy, "Travel and World Power: Americans in Europe, 1890–1917," *Diplomatic History* 22, no. 4 (2002): 565–94; Frank A Ninkovich, *The Wilsonian Century: U.S. Foreign Policy Since 1900* (Chicago: University of Chicago Press, 1999); McCormick, *America's Half-Century*.
98. Rostow, *The Process of Economic Growth*, 4, 6; Lerner, *The Passing of Traditional Society*, 71, 86.
99. Pye, *Guerrilla Communism in Malaya*; Robert Grainger Ker Thompson, *Defeating Communist Insurgency: Experiences from Malaya and Vietnam*, Studies in International Security, no. 10 (London: Chatto & Windus, 1966).
100. Pye, *Guerrilla Communism in Malaya*, 8, 348, 354.
101. See Cooper, *Colonialism in Question*, chap. 5.
102. See beginning of this chapter. Russell, "News Release by United States Embassy, Accra; September 11, 1961; Subject: Remarks of the American Ambassador Mr. Francis K. Russell at the Presentation of Books to the Ghana Military Academy."
103. Anderson, *Imagined Communities*, 120, 126.
104. Ibid., 116, 120.
105. Aboagye argues that Britain's exceptionally poor rate of training officers mitigates against thinking Nkrumah too rash in sacking them. Aboagye, *The Ghana Army*, 116.
106. Alexander, *African Tightrope*, 14; S. Kojo Addae, *History of Ghana Airforce*, vol. 4, *Military Histories* (Accra, Ghana: Eureka Publications, 2012), 28–29.
107. Aboagye, *The Ghana Army*, 379.

108. Ibid., 118.
109. Kwame Nkrumah, "Draft Speech by Kwame Nkrumah to Ghana Military Academy Passing Out Parade; September 29, 1962" (Ghana Military Academy, Teshie, September 29, 1962), RG 14/4/506, doc. 269, PRAAD.
110. Ibid.
111. Kwame Nkrumah, "Draft Speech by Kwame Nkrumah to Ghana Military Academy Passing Out Parade; September 24, 1963" (Ghana Military Academy, Teshie, September 24, 1963), RG 14/4/506, doc. 328, PRAAD.
112. G. D. Hunt, "Recollections of the Canadian Armed Forces Training Team in Ghana, 1961–1968," *Canadian Defense Quarterly* 18, no. 5 (April 1989): 46.
113. Adekson, "Pay, Promotion, and Other Self-Interests of Military Intervention in Politics"; Arlinghaus, *Military Development in Africa*, 13; Henry S. Bienen, "African Militaries as Foreign Policy Actors," in *Arms and the African: Military Influences on Africa's International Relations*, eds. William J. Foltz and Henry S. Bienen (New Haven, CT: Yale University Press, 1985), 153–70; Samuel Decalo, "Modalities of Civil-Military Stability in Africa," *Journal of Modern African Studies* 27, no. 4 (December 1, 1989): 547–78; Hutchful, "Institutional Decomposition"; Claude E. Welch Jr., "Civilian Control of the Military: Myth and Reality," in *Civilian Control of the Military: Theory and Cases from Developing Countries*, ed. Claude E. Welch Jr. (Albany: State University of New York Press, 1976), 1–42.
114. Frederick Cooper, "The Dialectics of Decolonization: Nationalism and Labor Movements in Postwar French Africa," in *Tensions of Empire: Colonial Cultures in a Bourgeois World*, eds. Frederick Cooper and Ann Laura Stoler (Berkeley: University of California Press, 1997), 406.
115. Ali A. Mazrui, "Soldiers as Traditionalizers: Military Rule and the Re-Africanization of Africa," *World Politics* 28, no. 2 (January 1, 1976): 246–72.
116. The phrase "double culture" first appears in Philip Mayer and Iona Mayer, *Townsmen or Tribesmen: Conservatism and the Process of Urbanization in a South African City*, Xhosa in Town 2 (Cape Town: Oxford University Press, 1961), 11; Amoah applies the term to urban Ghanaians who maintain their rural identities, complicating the "urbanization-as-detribilzation" dichotomy. Amoah, *Reconstructing the Nation in Africa*, 136.
117. Mayer and Mayer, *Townsmen or Tribesmen*, 10.
118. This is the consensus of participants and external scholars. See Afrifa, *The Ghana Coup, 24th February 1966*; Ocran, *Politics of the Sword*; Baynham, *The Military and Politics in Nkrumah's Ghana*; Kraus, "Arms and Politics in Ghana"; Price, "Military Officers and Political Leadership," 364; Adekson, "Pay, Promotion, and Other Self-Interests of Military Intervention in Politics," 16; Bennett suggests the 1966 coup may have had more

"ideological" motives than those that followed in "The Motivation for Military Intervention," 660.

119. Baynham, *The Military and Politics in Nkrumah's Ghana*, 134; William P. Mahoney Jr., "252. Telegram from the Embassy in Ghana to the Department of State, April 2, 1965," in *FRUS 1964–1968*, vol. xxiv, *Africa*, ed. Nina Davis Howland (Washington, DC: US Govt. Print. Off., 1995), 445.
120. Trimble, "237. Trimble to Williams, February 11, 1964"; "257. Memorandum for the Deputy Director of Central Intelligence (Helms), February 25, 1966," in *FRUS 1964–1968*, vol. xxiv, *Africa*, ed. Nina Davis Howland (Washington, DC: US Govt. Print. Off., 1995), 453–54; Robert W. Komer, "253. Memorandum from Robert W. Komer of the National Security Council Staff to the President's Special Assistant for National Security Affairs (Bundy); May 27, 1965," in *FRUS 1964–1968*, vol. xxiv, *Africa*, ed. Nina Davis Howland (Washington, DC: US Govt. Print. Off., 1995), 447.
121. Robert W. Komer, "260. Memorandum from the President's Acting Special Assistant for National Security Affairs (Komer) to President Johnson; March 12, 1966," in *FRUS 1964–1968*, vol. xxiv, *Africa*, ed. Nina Davis Howland (Washington, DC: US Govt. Print. Off., 1995), 457; Lt. Gen. J. A. Ankrah, "261. Letter from Chairman of the National Liberation Council Lieutenant General Ankrah to President Johnson, March 24, 1966," in *FRUS 1964–1968*, vol. xxiv, *Africa*, ed. Nina Davis Howland (Washington, DC: US Govt. Print. Off., 1995), 458–61.
122. Ankrah, "261. Ankrah to Johnson; March 24, 1966," 460; Nicholas deBelleville Katzenbach, "Secret Memorandum from Nicholas deBelleville Katzenbach, Undersecretary of State to the President, October 9, 1967; Subject: Your Meeting with General Ankrah," October 9, 1967, RG 59 CFPF 1967–1969, box 2140, folder POL 7 Ghana 1/1/67, doc. 17255, NACP; Franklin H. Williams, "Confidential Telegram 764 from Franklin Williams, Ambassador to Ghana, to Secretary of State; August 29, 1967" (American Embassy, Accra, August 29, 1967), RG 59 CFPF 1967–1969, box 2141, folder POL 23 Ghana 1/1/67, NACP; Dean Rusk, "Secret Telegram from Sec. of State Rusk to U.S. Embassy, Accra; September 11, 1967" (Washington, DC, September 11, 1967), RG 59 CFPF 1967–1969, box 2141; folder POL 23 Ghana 1/1/67, doc. 35324, NACP.
123. Benjamin H. Read, "Confidential Memorandum for Mr. Walt W. Rostow by Benjamin H. Read, Executive Secretary; Subject: Washington Visit of Ghana Head-of-State, General Joseph A. Ankrah; September 29, 1967," September 29, 1967, 3, RG 59 CFPF 1967–1969, box 2140, folder POL 7 Ghana 1/1/67, NACP.
124. Read, "Rostow to Read; Sept. 29, 1967; Subj: Ankrah Visit."

125. Dean Rusk, "Secret Telegram from Sec. of State Rusk to U.S. Embassy, Accra; May 1,1967; Subject: Ref: Telegrams Accra 2471 and London 7198" (Washington, DC, May 1, 1967), RG 59 CFPF 1967–1969, box 1690, folder DEF—Defense Affairs US-3 Ghana 1/1/67, doc. 185545, NACP.
126. Lyndon B. Johnson, "262. Letter from President Johnson to Chairman of the National Liberation Council Lieutenant General Ankrah, April 4, 1966," in *FRUS 1964–1968*, vol. xxiv, *Africa*, ed. Nina Davis Howland (Washington, DC: US Govt. Print. Off., 1995), 462; J. A. Ankrah, "Letter from Lt. Gen. J.A. Ankrah, Chairman of the National Liberation Council and Ghana Head of State, to Pres. Lyndon Johnson; June 27, 1967" (Accra, June 27, 1967), RG 59 CFPF 1967–1969, box 2140, folder POL 7 Ghana 1/1/67, NACP; Katzenbach, "Katzenbach to Johnson; Oct. 9, 1961; Subj: Meeting with Ankrah."
127. Katzenbach, "Katzenbach to Johnson; Oct. 9, 1961; Subj: Meeting with Ankrah."
128. Ankrah, "261. Ankrah to Johnson; March 24, 1966."
129. Johnson, "262. Johnson to Ankrah; April 4, 1966."
130. Afrifa, *The Ghana Coup, 24th February 1966*, 50.
131. Gutteridge, "The Political Role of African Armed Forces," 95.
132. Ibid., 103.
133. Ankrah, "Ankrah to Johnson; June 27, 1967."
134. Walt W. Rostow, "271. Memorandum From the President's Special Assistant (Rostow) to President Johnson; October 9, 1967," in *FRUS 1964–1968*, vol. xxiv, *Africa*, ed. Nina Davis Howland (Washington, DC: US Govt. Print. Off., 1995), 475–77.
135. Rusk, "Rusk to Accra; Sept. 11, 1961"; Rusk, "Secret Telegram from Sec. of State Rusk to U.S. Embassy, Accra; May 1, 1967; Subject: Ref: Telegrams Accra 2471 and London 7198."
136. Katzenbach, "Katzenbach to Johnson; Oct. 9, 1961; Subj: Meeting with Ankrah."
137. Read, "Rostow to Read; Sept. 29, 1967; Subj: Ankrah Visit"; Robert Dowse, "Military and Police Rule," in *Politicians and Soldiers in Ghana, 1966–1972*, eds. Dennis Austin and Robin Luckham (London: Cass, 1975), 16.
138. Franklin H. Williams, "Confidential Telegram from Franklin H. Williams, Ambassador to Ghana, to Department of State; August 10, 1967; Subject: Internal Security A Brief Status Report" (American Embassy, Accra, August 10, 1967), RG 59 CFPF 1967–1969, box 2141, folder POL 23 Ghana 1/1/67, NACP. Emphasis in original.
139. Ibid.

140. In 1966, $1.4 billion; $792 million in 1967; $500 million eventually appropriated in 1968; $375 million appropriated in 1969. DSAA, *Congressional Presentation, Security Assistance Programs*, FY 1981, 6.
141. Rusk, "Secret Telegram from Sec. of State Rusk to U.S. Embassy, Accra; May 1,1967; Subject: Ref: Telegrams Accra 2471 and London 7198."
142. Dean Rusk, "Confidential Telegram from Sec. of State Rusk to U.S. Embassy, Accra; October 12, 1967" (Washington, DC, October 12, 1967), RG 59 CFPF 1967–1969, box 2140, folder POL 7 Ghana 1/1/67, doc. 52667, NACP.
143. Katzenbach, "Katzenbach to Johnson; Oct. 9, 1961; Subj: Meeting with Ankrah." Emphasis in original.
144. Ibid.
145. Ibid.
146. Described in greater detail in Chapter 1. McNamara, "Draft Memorandum for the President on the Military Assistance Program, FY 1969–1973; December 18, 1967," 41.
147. Ibid.
148. Ibid., 12.
149. J. A. Ankrah, "Message from Lt-Gen J. A. Ankrah, OOV, MC, Chairman of the NLC and Commander-in-Chief of the Ghana Armed Forces," *Ghana Armed Forces Magazine*, February 1967, FES.
150. It is not possible to confirm whether the letters (some quite critical of the Ghana Armed Forces administration) were genuine.
151. Ankrah, "Message from Lt-Gen J. A. Ankrah, OOV, MC, Chairman of the NLC and Commander-in-Chief of the Ghana Armed Forces," 3.
152. Ibid.
153. The presence of copies of this magazine at the Friedrich-Ebert-Stiftung Library in Bonn, Germany, and their absence from the Library of Congress betrays some of the international circulation of the first several issues of *Ghana Armed Forces Magazine*. Friedrich-Ebert-Stiftung Library, which specializes in German and international labor movements, only held issues one through six (February 1967 through October 1969). Library of the Friedrich Ebert Foundation home page, accessed September 10, 2013, *library.fes.de/index_en.htm*. On the other hand, the director of public relations for the Ghana Armed Forces did not have copies of these six editions.
154. The Staff, "The Rationale Behind Military Intervention in Politics," *Ghana Armed Forces Magazine*, June 1969, 21, FES. About half the magazine's articles listed their authors by name. Many, like this one, list a pen name.
155. Ghanaian sources, especially military ones, frequently use only first and middle initials with last names. In some cases, we can find their first names

elsewhere, but *Armed Forces News* rarely published them. We will see first initials regularly in this chapter and Chapter 3. Col. A. K. Kattah, "Foundations of Defense Policy," *Ghana Armed Forces Magazine*, February 1968, 14, FES.

156. Capt. S. G. Amoo, "Selling the Forces to the Taxpayer," *Ghana Armed Forces Magazine*, February 1968, 27, FES It's a reasonable suspicion, although I have no evidence, that Amoo first read this quote as a student at a public affairs officers school in the United States.
157. For more on the stigma, see also Afrifa, *The Ghana Coup, 24th February 1966*, 93.
158. Amoo, "Selling the Forces to the Taxpayer," 27.
159. A Staff Correspondent, "'Operation Yellow Pod': Armed Forces Cart Cocoa," *Ghana Armed Forces Magazine*, June 1969, FES.
160. Lt. A. Enninful, "The Change from Civil to Military," *Ghana Armed Forces Magazine*, February 1967, FES.
161. Ibid., 19–20. Transcribed exactly from original.
162. Ibid. Asafos were Akan warrior groups in the seventeenth and eighteenth centuries.
163. Ibid., 20. "Corrupt" is a play on Nkrumah's Convention People's Party (CPP).
164. Ibid.
165. Air Marshal M. A. Otu, "GOC On Ghana's Defence Policy," *Ghana Armed Forces Magazine*, June 1968, 10, FES.
166. Ibid., 11.
167. Ibid. Verbatim from text.
168. Ibid. The term "men" refers to the enlisted members of the military (not officers). The phrase "other ranks" commonly appears as well.
169. Ibid. Verbatim from text.
170. Ibid.
171. See Hutchful, "Institutional Decomposition," 212.
172. Enninful, "The Change from Civil to Military," 20.
173. Tachie-Menson, "Education of Military Leaders," 10.
174. Ibid.
175. "Military Personnel and Politics," *Ghana Armed Forces Magazine*, June 1969, FES.
176. Ibid.
177. Frederick S. Arkhurst, "A Brief Survey of Disarmament Negotiations," in *Arms and African Development: Proceedings of the First Pan-African Citizens' Conference*, ed. Frederick S. Arkhurst, (New York: Praeger, 1972), 18.
178. Frederick S. Arkhurst, ed., *Arms and African Development: Proceedings of the First Pan-African Citizens' Conference*, (New York: Praeger, 1972), ix.

179. Ibid., x.
180. Kraus, "Arms and Politics in Ghana," 217; Welch, "The African Military and Political Development," 228.
181. Huntington, *Political Order in Changing Societies*, 194.
182. Valerie Plave Bennett, "Epilogue: Malcontents in Uniform," in *Politicians and Soldiers in Ghana, 1966–1972*, eds. Dennis Austin and Robin Luckham (London: Cass, 1975), 660; Austin and Luckham, *Politicians and Soldiers in Ghana, 1966–1972*, 3; Decalo quotes Ruth First in Samuel Decalo, *Coups and Army Rule in Africa: Studies in Military Style* (New Haven, CT: Yale University Press, 1976), 19; Ruth First, *Power in Africa*, 1st American ed. (New York: Pantheon Books, 1970), 20; Simon Baynham, "Divide et Impera: Civilian Control of the Military in Ghana's Second and Third Republics," *Journal of Modern African Studies* 23, no. 4 (December 1, 1985): 628.
183. Embassy of Ghana, "Dr. Busia at the National Press Club," *Ghana News*, December 1969.
184. "Transcript of Televised Interview Granted to Mr. Carl Rowan, Former U.S. Ambassador to Finland and Director of U.S. Information Agency by the Rt. Hon. Dr. K. A. Busia, Prime Minister at the Castle [Otu] on Friday, 14th August, 1970" (Transcript, Accra, August 14, 1970), RG 59 SNF 1970–1973, box 2323, folder POL 15-1 Ghana, 1/1/71, NACP.
185. Embassy of Ghana, "Apartheid in South Africa: Dr. Busia Calls for Additional Methods"; William P. Rogers, "Confidential Memorandum from Secretary of State William P. Rogers to President Johnson, November 2, 1971; Subject: Call by Prime Minister Busia of Ghana November 4 (1971), 4:00 Pm" (Washington, DC, November 2, 1971), RG 59 SNF 1970–1973, box 2322, folder POL 7 Ghana, NACP. For US policy shift toward accommodating and "constructive engagement" with South Africa, see Irwin, *Gordian Knot*; Thomas Borstelmann, *Apartheid's Reluctant Uncle: The United States and Southern Africa in the Early Cold War* (New York: Oxford University Press, 1993).
186. Embassy of Ghana, "Apartheid in South Africa: Dr. Busia Calls for Additional Methods"; Rogers, "Confidential Memorandum from Secretary of State William P. Rogers to President Johnson, November 2, 1971; Subject: Call by Prime Minister Busia of Ghana November 4 (1971), 4:00 Pm."
187. "Transcript of Televised Interview Granted to Mr. Carl Rowan, Former U.S. Ambassador to Finland and Director of U.S. Information Agency by the Rt. Hon. Dr. K. A. Busia, Prime Minister at the Castle [Otu] on Friday, 14th August, 1970," 8.
188. J. H. McFarland Jr., "Secret Telegram from J. H. McFarland, U.S. Embassy, Accra to Department of State, June 15, 1972; Subject: Biographic

Information on Newly Appointed Ghanaian Ambassadors" (Memorandum, Accra, June 15, 1972), RG 59 SNF 1970–1973, box 2322, folder POL 7 Ghana, NACP; Bennett, "Epilogue: Malcontents in Uniform," 309.

189. NSC Interdepartmental Group for Africa, "Secret Ghana Policy Planning Paper by NSC Interdepartmental Group for Africa, May 1971," 2. Author assumes "tying," not "trying" was intended.
190. Ibid., 6.
191. Ibid., 2.
192. Ibid., 6.
193. Ibid., 7.
194. J. H. McFarland Jr., "Confidential Telegram from J. H. McFarland, Jr., US Embassy Ghana to Department of State, December 31, 1970; Subject: Ghana Draft Country Policy Paper" (Policy Paper, Accra, December 31, 1970), 4, RG 59 SNF 1970–1973, box 2325, folder POL 1 Ghana-US 10/15/70, NACP. Emphasis is mine.
195. Fred L. Hadsel, "Secret Telegram from U.S. Ambassador Fred L. Hadsel to the Secretary of State, January 20, 1972; Subject: Ghana Coup" (Telegram, Accra, January 20, 1972), RG 59 SNF 1970–1973, box 2325, folder POL 23-9 Ghana 1/1/72, NACP. This telegram clips many articles and pronouns. I add them in the text for readability, hoping that will distract the reader less than their absence might.
196. Ibid.
197. Fred L. Hadsel, "Secret Telegram from U.S. Ambassador Fred L. Hadsel to U.S. State Department, January 14, 1972; Subject: Ghana Coup: Biodata on NRC Members" (Telegram, Accra, January 14, 1972), RG 59 SNF 1970–1973, box 2324, folder POL 23-9 Ghana, 1/1/72, doc. Accra 309, NACP.
198. Ibid.
199. Ibid. In an unusual point of style, Hadsel's cable includes no ranks. This is also different from most Ghanaian printed sources, which rarely include first names in even the most sympathetic references to military figures; Bennett, "Epilogue: Malcontents in Uniform," 306.
200. Fred L. Hadsel, "Confidential Telegram from U.S. Ambassador Fred L. Hadsel to U.S. State Department, February 26, 1972; Subject: Defense Attache Courtesy Call on Chairman Acheampong" (Telegram, Accra, February 26, 1972), RG 59 SNF 1970–1973, box 2323, folder POL 15 Ghana, 1/1/70, doc. Accra 12147, NACP.
201. Ibid.
202. Ibid.; Fred L. Hadsel, "Secret Telegram from U.S. Ambassador Fred L. Hadsel to the Secretary of State, January 17, 1972; Subject: Relations with New Ghana Regime" (Telegram, Accra, January 17, 1972), RG 59 SNF

1970–1973, box 2323, folder POL 15 Ghana 1/1/72, doc. Accra 360.1-1972, NACP.

203. Bennett, "Epilogue: Malcontents in Uniform," n. 11.
204. "General Kutu A. Acheampong: The Fifth Milestone; 13th January 1976–12th January 1977" (Accra: The Office of the Press Secretary to the Supreme Military Council, 1977), 286, George Padmore Research Library, Accra.
205. "Transcript of an Interview Granted by Colonel I. K. Acheampong, Chairman, National Redemption Council to Anthony Hugues, Editor of 'Africa Magazine,' Paris—April, 1972," in *Speeches and Interviews by Col. I. K. Acheampong*, vol. 1 (Accra, 1972), 89; "Transcript of an Interview with the Head of State, Colonel I. K. Acheampong, Chairman, National Redemption Council by Mr. Harri Czepuck, Correspondent of the 'Horixont,' a Weekly Publication in the German Democratic Republic, February 23, 1973," in *Speeches and Interviews by Col. I. K. Acheampong*, vol. 2 (Accra, 1973), 3.
206. "Interview with Acheampong by Czepuck, February 23, 1973," 3; "Interview with Acheampong by Hugues, April 1972," 87.
207. "Interview with Acheampong by Czepuck, February 23, 1973"; Bennett, "Epilogue: Malcontents in Uniform," 309.
208. "Interview with Acheampong by Hugues, April 1972," 89.
209. Ignatius Kutu Acheampong, "The Charter—A Guide to Redemption (Part 1)," *Ghana Armed Forces News*, September 1, 1974.
210. "Interview with Acheampong by Hugues, April 1972," 86.
211. Tom Dorkenoo, "Book Review: Politicians and Soldiers in Ghana, 1966–1972," *Ghana Armed Forces News*, May 15, 1976, GAF DPR; Bennett, "Epilogue: Malcontents in Uniform," cited above.
212. The SMC, Supreme Military Council, replaced Acheampong's junta, the NRC, in 1975. Dorkenoo, "Book Review: Politicians and Soldiers in Ghana, 1966–1972."
213. Henry A. Kissinger, "Confidential Telegram from Secretary of State to All American Diplomatic Posts, 3 March 1976, Subject: Changes in Foreign Military Training Program" (Telegram, Washington, DC, March 3, 1976), RG 59 CFPF 1973–1976, doc. 1976State051768 in Electronic Telegrams, 1/1/1976—12/31/1976, NACP.
214. Shirley Temple Black, "Confidential Telegram from Shirley Temple Black, US Ambassador to Ghana, to Secretary of State; 12 March 1976, Subject: Changes in Pricing of Foreign Military Training Program" (Telegram, Accra, Ghana, March 12, 1976), RG 59 CFPF 1973–1976, doc. 1976Accra01906 in Electronic Telegrams, 1/1/1976—12/31/1976, NACP.
215. Ibid. Original text says, "his possible assumptiokntln key roles either in" Author assumes this is a typographical error.

216. Ibid.
217. Ibid.
218. Shirley Temple Black, "Confidential Telegram from Ambassador Shirley Temple Black to Secretary of State Kissinger, April 27, 1976; Subject: Cancellation of Secretary's Visit" (Telegram, Washington, DC, April 27, 1976), RG 59 CFPF 1973–1976, doc. 1976State051768 in Electronic Telegrams, 1/1/1976—12/31/1976, NACP.
219. According to Addae, Ghana's standards mirrored other Western militaries regarding staff college: junior officers must complete a primary or junior staff college prior to promotion to major. Usually an officer could not progress beyond the rank of major without completing a senior staff college course. Ft. Leavenworth's Command and General Staff College and the Ghana Armed Forces Staff College represented this type of training. Addae, *History of Ghana Armed Forces: Military Organizations and Regiments*, 3:253.
220. Ibid.
221. Nkrumah, *Africa Must Unite*, 220; Research and Development Department, *Ghana Armed Forces Staff College*, 9.
222. Research and Development Department, *Ghana Armed Forces Staff College*, 9.
223. Addae, *History of Ghana Armed Forces*, 3:254–255.
224. Research and Development Department, *Ghana Armed Forces Staff College*, 15.
225. Ibid., 33–34.
226. Ibid., 18.
227. Ibid., 84–102.
228. Ibid., 38.
229. "Address by the Head of State and Chairman of the NRC [Acheampong] on the Occasion of the Passing Out Parade for Intake Number Sixteen—Ghana Military Academy—15th March, 1975" (Speech, Ghana Military Academy, Teshie, March 15, 1975), 251, RG 14/4/507 page 245–251, PRAAD.
230. Research and Development Department, *Ghana Armed Forces Staff College*, 19.
231. Ibid., 31.
232. Messages from the United Nations tasking Ghana to fill staff positions specifically requested "staff trained" officers. "Message from Secretary General's Office to Permanent Representative of Ghana to the United Nations; 29 May 1981; Subject: Staff Officer Requirements" (New York, May 29, 1981), AG-20; S-0356 UNIFIL Subject Files, box 11, folder 6, UN ARMS; see also box S—0332—007 file 15 for a 1976 request. Maj. A. A. Enninful, "To the War Zone and Back," *Ghana Armed Forces News*, July 31, 1974, GAF DPR.

233. Achel, interview with author.
234. Actually, there were nine men on the team, but only seven served as faculty, including the deputy commandant. Research and Development Department, *Ghana Armed Forces Staff College*, 30.
235. Ibid., 31.
236. "Personalities," *The Flame: The Magazine of the GAFCSC*, 1988, GAFCSC Library.
237. Achel, interview with author.
238. Research and Development Department, *Ghana Armed Forces Staff College*, 41. The junior division of the staff college remained open, as did some institutes that trained civilian personnel. In order for my terminology to match between American staff colleges (that generally serve midgrade officers like majors and lieutenant colonels), I maintain the term "staff college" to refer only to the senior division of Ghana Armed Forces Staff College.
239. Achel, interview with author.
240. Ibid.
241. Michael Abana, interview with author, audio recording, July 18, 2012, author's possession.
242. Research and Development Department, *Ghana Armed Forces Staff College*, 104.
243. Abana, interview with author; Research and Development Department, *Ghana Armed Forces Staff College*, 22.
244. Moya, *Cousins and Strangers*, 4.
245. Ibid., 394.
246. See Ngugi wa Thiongo, *Decolonising the Mind*.

## CHAPTER 3

Portions of this research appear in "UN Peacekeeping and the International Men and Women of the Ghana Armed Forces." *International Journal of Military History and Historiography* 36, no. 1 (June 28, 2016): 9–33. Reprinted with permission.

1. Lt. Makana N. Blay, "From Makana's Fanara Diary," *Ghana Armed Forces News*, February 15, 1975, 4, GAF DPR.
2. The cedi is Ghana's national currency. The "Kissinger" is named after US Secretary of State Henry Kissinger, who helped negotiate the 1973 cease-fire and UN mission in Sinai.
3. Capt. Georgina Mensah became the first Ghanaian woman peacekeeper in 1985. Stephen Owusu, "The Lone GHANBATT Lady," *Litani: UNIFIL News*, February 1985, 5, AG-20, S-1078 Personal Files, box S-1078-0089, folder 5—Litani—UNIFIL News Jan–Jun 1985, UN ARMS.
4. See Akyeampong, "'Diasporas,' Mobility and the Social Imaginary."

5. Anderson, *Imagined Communities*; Manger and Assal, "Diasporas Within and Without Africa," 16; Patrick Manning, *The African Diaspora: A History through Culture* (New York: Columbia University Press, 2009), 323; Robert Young, *Postcolonialism: A Very Short Introduction* (Oxford: Oxford University Press, 2003), 61; Amoah, *Reconstructing the Nation in Africa*, 4.
6. Korang, *Writing Ghana, Imagining Africa*, 12; Amoah, *Reconstructing the Nation in Africa*, 5.
7. Baynham calls this a "fragmented" rather than "integral" relationship with their political and social environments. Simon Baynham, "Introduction: Armed Forces in Africa," in *Military Power and Politics in Black Africa*, ed. Simon Baynham (London: Croom Helm, 1986), 7.
8. Hutchful, "Institutional Decomposition," 214.
9. Ibid., 251.
10. Decalo, *Coups and Army Rule in Africa*, 56; Herbert M. Howe, *Ambiguous Order: Military Forces in African States* (Boulder, CO: Lynne Rienner Publishers, 2001), 57; Huntington, *The Soldier and the State*, chap. 4; Huntington, "Political Development and Political Decay," 429; Eboe Hutchful, "Military Policy and Reform in Ghana," *Journal of Modern African Studies* 35, no. 2 (June 1, 1997): 252.
11. Eboe Hutchful, "A Tale of Two Regimes: Imperialism, the Military and Class in Ghana," *Review of African Political Economy*, no. 14 (January 1, 1979): 41.
12. Including: Namibia, Angola, Western Sahara, Somalia, Mozambique, Rwanda, Chad, Afghanistan, Pakistan, Cambodia, Bosnia-Herzegovina, Croatia, Macedonia, Slavonia, Iran, Iraq, Kuwait, East Timor, Kosovo, Liberia, and Sierra Leone. The number 80,000 includes repeats—it is a tally of participants, not individual soldiers. United Nations Information Center, *Ghana: Four Decades of UN and Regional Peacekeeping* (Accra, Ghana: United Nations Information Center, 2000), 26–27.
13. Ghanaian women soldiers began peacekeeping in very small numbers in 1985; by the 1990s, they participated in larger numbers. Major Comfort Ankomah-Danso describes her experiences briefly in "Breaking the Barriers: Women in Peacekeeping" in ibid., 18; see also Owusu, "The Lone GHANBATT Lady," 5.
14. Anderson's "imagined community" is the nation-state. Iriye's "imagined world" is global. Anderson, *Imagined Communities*; Iriye, "Internationalizing International History," 53.
15. Akyeampong, "Africans in the Diaspora," 183; Manuh, "'Efie' or the Meanings of 'Home,'" 143; Manning, *The African Diaspora*, xvii; William Ackah, "The Intersection of African Identities in the Twenty-First Century: Old and New Diasporas and the African Continent," in *Reframing*

*Contemporary Africa: Politics, Economics, and Culture in the Global Era*, eds. Peyi Soyinka-Airewele and Rita Kiki Edozie (Washington, DC: CQ Press, 2010), 131; Arthur, *African Diaspora Identities*, xii; Kamari Maxine Clarke, "New Spheres of Transnational Formations: Mobilizations of Humanitarian Diasporas," *Transforming Anthropology* 18, no. 1 (2010): 48.

16. Akyeampong, "'Diasporas,' Mobility and the Social Imaginary," 25.
17. Although, Abaka maintains that "Hausa" identity persisted, not that a "military" identity replaced it. Edmund Abaka, "Traders, Slaves, and Soldiers: The Hausa Diaspora in Ghana (Gold Coast and Asante) in the Nineteenth and Early Twentieth Centuries," in *Movements, Borders, and Identities in Africa*, eds. Toyin Falola and Aribidesi Adisa Usman (Rochester, NY: University of Rochester Press, 2009), 194; Arthur, *African Diaspora Identities*, 128; Manger and Assal, "Diasporas Within and Without Africa," 17; Paul Gilroy, *The Black Atlantic: Modernity and Double Consciousness* (Cambridge, MA: Harvard University Press, 1993), 128.
18. Baynham, "Divide et Impera"; Hutchful, "Military Policy and Reform in Ghana"; Dennis Austin, "The Ghana Armed Forces and Ghanaian Society," *Third World Quarterly* 7, no. 1 (January 1, 1985): 90–101; Addae, *A Short History of Ghana Armed Forces*, 137; Baynham, *The Military and Politics in Nkrumah's Ghana*; Aboagye, *The Ghana Army*; Edward Feit, "Pen, Sword, and People: Military Regimes in the Formation of Political Institutions," *World Politics* 25, no. 2 (January 1, 1973): 251–73.
19. Addae, *A Short History of Ghana Armed Forces*. See also Chazan, *An Anatomy of Ghanaian Politics*.
20. Addae employs the common practice of using only first initials when discussing minor military officials in historical work. Sandhurst's records do not include his first name, which is protected by the British Data Protection Act. A. R. Morton, "RE: Request for Research Assistance Regarding Ghanaian Cadets at RMA Sandhurst," January 18, 2012, author's possession.
21. Ibid.
22. Addae, *History of Ghana Armed Forces*, 3:74.
23. Ibid., 3:76.
24. Ibid.
25. Anyidoho, *My Journey . . . Every Step*, 19.
26. Enninful, "The Change from Civil to Military," 20.
27. Akyeampong, "'Diasporas,' Mobility and the Social Imaginary," 30.
28. Black, "Black to Kissinger; Mar. 12, 1976; Subj: Changes in Pricing of Military Training"; Dan Henk, *Uncharted Paths, Uncertain Vision U.S. Military Involvements in Sub-Saharan Africa in the Wake of the Cold War*, Occasional Paper 18 (Colorado Springs, CO: USAF Institute for National

Security Studies, US Air Force Academy, 1998), 29; William H. McCoy, *Senegal and Liberia: Case Studies in U.S. IMET Training and Its Role in Internal Defense and Development*, N-3637-USDP (Santa Monica, CA: RAND, 1994), 4.

29. Air force commander, army commander, etc. were equivalent to the American rank of chief of staff. Bennett, "Epilogue: Malcontents in Uniform," 304.
30. Fred L. Hadsel, "Telegram from U.S. Ambassador Fred L. Hadsel, Ghana to U.S. State Department, February 10, 1972; Subject: Changes in NRC" (Telegram, Accra, February 10, 1972), RG 59 SNF 1970–1973, box 2323, folder POL 15 Ghana, 1/1/70, NACP.
31. Achel, interview with author.
32. Anyidoho, *My Journey . . . Every Step*, 64.
33. Ibid.
34. Oduro-Kwarteng, *The Memoirs of a Colonel Retired*, 40; Baynham, "Divide et Impera," 636.
35. Aboagye, *The Ghana Army*, 103.
36. Anyidoho, *My Journey . . . Every Step*, 65.
37. Ibid.
38. Ibid.
39. Addae, *History of Ghana Armed Forces*, 3:26.
40. Ibid., 3:26, 143.
41. Anyidoho, *My Journey . . . Every Step*, 65.
42. Ibid., 26.
43. Ibid., 27.
44. Ibid., 60.
45. Ibid., 61.
46. Ibid.
47. Ibid., 62.
48. Arthur, *African Diaspora Identities*, 51–69.
49. Achel, interview with author.
50. Anyidoho, *My Journey . . . Every Step*, 61.
51. Arthur, *African Diaspora Identities*, 79.
52. Manuh derives precisely the same conclusions observing Ghanaian families in Toronto, Canada. Manuh, "'Efie' or the Meanings of 'Home,'" 157.
53. This reflects another of Manuh's observations. Ibid.
54. Arthur, *African Diaspora Identities*, 132.
55. Akyeampong, "Africans in the Diaspora," 185; Manuh, "'Efie' or the Meanings of 'Home,'" 146; Owusu, "The Role of Ghanaian Immigrant Associations in Toronto, Canada," 1174–1175; Arthur, *African Diaspora Identities*, 138.

56. Mann considers this ability or need to remain "present" in home communities an important concern for West Africans serving in French colonial armies. It matches, though not precisely, similar arguments from Manuh. Mann, *Native Sons*, 177.
57. Anyidoho, *My Journey . . . Every Step*, 65.
58. Ibid., 66.
59. Ibid.
60. Ibid., 67.
61. Ibid., 66.
62. Ibid., 69.
63. Ibid.
64. Abana, interview with author.
65. Ibid.
66. Ibid.
67. Ibid.
68. Oduro-Kwarteng, *The Memoirs of a Colonel Retired*, 34. Although neither Oduro-Kwarteng nor Anyidoho uses names, they specifically dismiss the events as regrettable and state they were both mistakenly accused. Anyidoho probably thought Oduro-Kwarteng was the "junior signals officer" who sabotaged his Fort Leavenworth course. Oduro-Kwarteng tried to defend himself from Anyidoho's accusation in his book. It is curious that the two most detailed memoirs by Ghanaian military officers should specifically clash with one another.
69. Ibid., 43.
70. Ibid., 39.
71. Ibid., 41.
72. Baynham, "Divide et Impera," 637.
73. Ibid., 638. Oduro-Kwarteng does not mention if his three-year academic posting to India in September 1979 was a politically motivated reward.
74. Oduro-Kwarteng, *The Memoirs of a Colonel Retired*, 43.
75. Ibid., 44.
76. Ibid., 45.
77. Ibid., 46.
78. Ibid.
79. Appiah, *In My Father's House*, 168.
80. Arthur, *African Diaspora Identities*, 128.
81. Ibid.; Manuh, "'Efie' or the Meanings of 'Home,'" 157.
82. Brian Urquhart, "Code Cable from Urquhart to Erskine; August 16, 1979; Subject: Ghana Contingent in UNIFIL" (New York, August 16, 1979), AG-20; S-0356 UNIFIL Subject Files, box 11, folder 6, UN ARMS;

Emmanuel A. Erskine, "Code Cable from Erskine to Urquhart; August 11, 1979; Subject: Planning for Employment of GHANBATT in UNIFIL" (New York, August 11, 1979), AG-20; S-0356 UNIFIL Subject Files, box 11, folder 6, UN ARMS; Brian Urquhart, "Code Cable from Urquhart to Siilasvuo and Erskine; August 10, 1979; Subject: Siilasvuo's Meeting with Eitan" (New York, August 10, 1979), AG-20; S-0356 UNIFIL Subject Files, box 46, folder 8, UN ARMS.

83. Erskine, *Mission with UNIFIL*, 155.
84. Hadsel, "Telegram from U.S. Ambassador Fred L. Hadsel, Ghana to U.S. State Department, February 10, 1972; Subject: Changes in NRC."
85. Hannes Philipp, "Autobriography [sic] of Brigadier-General Erskine" (New York, February 28, 1975), AG-20; S-0332 UNEF Subject Files, box S-0332-007, folder 6 Ghana 1973–1979, UN ARMS.
86. Erskine, *Mission with UNIFIL*, 160.
87. Maj. Gen. Odartey-Wellington died in action on June 4, 1979; Rawlings executed Akuffo and Kotei. Aboagye, *The Ghana Army*, 103, 390.
88. Emmanuel A. Erskine, "Code Cable from Erskine to Urquhart; June 25, 1979; Subject: Visit to Nigeria" (Naqoura, Lebanon, June 25, 1979), AG-20; S-0356 UNIFIL Subject Files, box 32, folder 1, UN ARMS.
89. Ensio Siilasvuo, "Incoming Code Cable from Siilasvuio to Urquhart/Guyer; September 20, 1974, Subject: Urquharts Cable 1529 to Guyer (Erskine Pay)" (New York, September 20, 1974), AG-20; S-0332 UNEF Subject Files, box S-0332-007, folder 7 Ghana 12/111973-11/11/1974, UN ARMS; "Biographical Note; Major-General Emmanuel Alexander Erskine" (New York, December 22, 1975), AG-20; S-0332 UNEF Subject Files, box S-0332-007, folder 6 Ghana 1973–1979, UN ARMS.
90. "Top Post for Erskine," *Ghana Armed Forces News*, December 15, 1974, 6, GAF DPR; Embassy of Ghana, "Ghana News, Vol 6 No 7 1974" (Washington, DC: Embassy of Ghana, April 30, 1974), 4–5; "Erskine Rides High," *Ghana Armed Forces News*, September 15, 1975, 1, GAF DPR; "Gen. Erskine Takes On New Job," *Ghana Armed Forces News*, March 15, 1976, 5, GAF DPR; "UNTSO in Close Ties With UNEF," *Ghana Armed Forces News*, September 15, 1976, 11; Embassy of Ghana, "Erskine: Army Must Be Loyal to Govt.," *Ghana News*, February 1981, 5.
91. Anyidoho, *My Journey . . . Every Step*, 68. Unlike Fort Leavenworth students, the Anyidohos had only a "military sponsor" and not a separate sponsor family (or two) from the civilian community around the base.
92. Ibid.
93. Henry Kwami Anyidoho, *Guns Over Kigali* (Accra, Ghana: Woeli Publishing Services, 1997), xiii; Anyidoho, *My Journey . . . Every Step*, 69.
94. Anyidoho, *My Journey . . . Every Step*, 69.

95. "Letter from F. E. Boaten, Ghana Ambassador to the United Nations to L. A. Koho, United Nations Secretariat; March 25, 1974; Subject: Brigadier Erskine's Curriculum Vitae" (New York, March 25, 1974), AG-20 UN Office for Special Political Affairs; S-0332 UNEF Subject Files, box S-0332-007, folder 10 Countries—Ghana [UNEF II] 4/1/1978-17/8/1979, UN ARMS.
96. Erskine, *Mission with UNIFIL*, 1.
97. Baynham, *The Military and Politics in Nkrumah's Ghana*, 99.
98. Ibid., 120.
99. See also Hutchful, "Institutional Decomposition," 212.
100. Among others, see "Message from Secretary General's Office to Permanent Representative of Ghana to the United Nations; 29 May 1981; Subject: Staff Officer Requirements"; "Memorandum from UN Secretary General's Office to Permanent Mission of Ghana; Subject: UNEF Headquarters Supplemental Personnel; February 1, 1974" (New York, February 1, 1974), AG-20; S-0332 UNEF Subject Files, box S-0332-007, folder 6 Ghana 1973–1979, UN ARMS.
101. The groups of Ghanaians were called a battalion and identified sequentially (i.e., GHANBATT 2 succeeded GHANBATT 1 in August 1974). These battalion designations are unrelated to the Ghana Army's six infantry battalions. "Memorandum from UN Secretary General's Office to Permanent Mission of Ghana; Subject: UNEF Headquarters Supplemental Personnel; February 1, 1974"; "Message from Secretary General's Office to Permanent Representative of Ghana to the United Nations; 29 May 1981; Subject: Staff Officer Requirements"; Aboagye, *The Ghana Army*, 155, 331.
102. See Arturo C. Sotomayor, *The Myth of the Democratic Peacekeeper: Civil-Military Relations and the United Nations* (Baltimore: John Hopkins University Press, 2014), 2.
103. Kwesi Aning, "Unintended Consequences of Peace Operations for Troop-Contributing Countries from West Africa: The Case of Ghana," in *Unintended Consequences of Peacekeeping Operations*, eds. Chiyuki Aoi, Cedric De Coning, and Ramesh Chandra Thakur (New York: United Nations University Press, 2007), 134–136.
104. Kwesi Aning and Festus K. Aubyn, "Ghana," in *Providing Peacekeepers: The Politics, Challenges, and Future of United Nations Peacekeeping Contributions*, eds. Alex J. Bellamy and Paul Williams (Oxford: Oxford University Press, 2013), 288–289; Aning, "Unintended Consequences of Peace Operations for Troop-Contributing Countries from West Africa: The Case of Ghana," 138; Abana, interview with author.
105. "Specialists" in some skills received extra pay, with limits. Special Committee On Peace-Keeping Operations, "Material On United Nations Observer Missions and Peace-Keeping Forces Authorized by the Security Council"

April 11, 1977, 47, AG-035; S-0531 General Records-Central Registry Finance Series, box 08, folder 4, UN ARMS.

106. In nominal US dollars. "UNEF Financing: Discussion Paper for Top Staff Meeting; April 17, 1974" (New York, April 17, 1974), AG-31; S-0534 Administrative Files 1948–1987, box 0448, folder S-1737-0080, UN ARMS; Special Committee On Peace-Keeping Operations, "Material On United Nations Observer Missions and Peace-Keeping Forces Authorized by the Security Council," 47; Patricio Ruedas, "Memorandum from Patricio Ruedas, Assistant Sec-Gen for Financial Services to Sec-Gen Waldheim; October 22, 1980; Subject: UNDOF/UNIFIL—Standard Rates of Reimbursement to Troop Contributors for Pay and Allowances of Their Troops" (New York, October 22, 1980), AG-006; S-0899 Peace-Keeping—Middle East, box S-0899-05, folder 11: Peacekeeping-UNIFIL-Meetings with Troop Contributors; "Message from the Permanent Mission of Ghana to the UN Secretary General; April 14, 1983; Subject: Reimbursement of Claims for Pay and Allowances and Equipment Depreciation" (New York, April 13, 1983), AG-20; S-0356 UNIFIL Subject Files, box 11, folder 7, UN ARMS; Erskine, *Mission with UNIFIL*, 157; Robert C. R. Siekmann, *National Contingents in United Nations Peace-Keeping Forces* (Dordrecht, Netherlands: Martinus Nijhoff, 1991), 165.

107. Thirty-one thousand is the total number of deployed persons, not separate individuals. Obviously, many soldiers deployed more than once, and each tour counts toward the total in this tally. Aboagye, *The Ghana Army*, 119. Most peacekeepers were army personnel, but not all. Addae, *History of Ghana Armed Forces*, 3:3. In total, 8,862 Ghanaians served in UNEF II; 45,500 served in UNIFIL between 1979 and 2000. Given the steady size of the Ghana detachment, I interpolated that that to roughly 23,000. United Nations Information Center, *Ghana: Four Decades*, 7, 26.

108. Special Committee On Peace-Keeping Operations, "Material On United Nations Observer Missions and Peace-Keeping Forces Authorized by the Security Council," 41.

109. "Message from the Secretary-General to the Permanent Representative of Ghana; Subject: Increasing Size of Ghana Contingent; March 12, 1982" (New York, March 12, 1983), AG-20; S-0356 UNIFIL Subject Files, box 11, folder 6, UN ARMS.

110. This is a complicated calculation, using Ghana's UN missions reimbursement claims from 1980 to 1982 (US$3.1 to $3.9 million), converted into real dollars for the year 2000 to match with the first available military expenditure data for Ghana (1984) in Omitoogun. "Message from the

Permanent Mission of Ghana to the UN Secretary General; April 14, 1983; Subject: Reimbursement of Claims for Pay and Allowances and Equipment Depreciation"; Wuyi Omitoogun, *Military Expenditure Data in Africa: A Survey of Cameroon, Ethiopia, Ghana, Kenya, Nigeria and Uganda*, Stockholm International Peace Research Institute (SIPRI) Reports 17 (Oxford: Oxford University Press, 2003), 132; "Inflation Calculator," Bureau of Labor Statistics," accessed March 5, 2014, *www.bls.gov/data/inflation_calculator.htm*; Hutchful corroborates these numbers, roughly, stating that Ghana billed the UN US$5.1 million per six-month rotation in Lebanon, of which 3.5 million went to troops and the government kept 1.6 million. Hutchful, "Military Policy and Reform in Ghana," 270.

111. "UNEF—I and II Mandates (for the Period 25 October 1973—24 October 1974)" (New York, October 1974), AG-31; S-0534 Administrative Files 1948–1987, box 48, folder S-1737-80, UN ARMS; "Info on UNEF & UNDOF Strength and Rotations" August 21, 1974, AG-31; S-0534 Administrative Files 1948–1987, box 48, folder S-1737-0104, UN ARMS; Brian Urquhart, "Memorandum from Brian Urquhart (Under-Sec-Gen for Special Pol. Affairs) to Alice Weil (Asst Sect-Gen, Office of General Services); November 5, 1984; Subject: Request from the Government of Ghana for the Use of Ghana Airways to Rotate the Ghana Contingent of UNIFIL" (New York, November 5, 1984), AG-20; S-0356 UNIFIL Subject Files, box 11, folder 7, UN ARMS; Erskine, *Mission with UNIFIL*, 157.
112. Embassy of Ghana, "Erskine: Army Must Be Loyal to Govt.," 5.
113. Manuh, "'Efie' or the Meanings of 'Home,'" 142; Akyeampong, "Africans in the Diaspora," 207. See also Mann, *Native Sons*, 177.
114. Oduro-Kwarteng, *The Memoirs of a Colonel Retired*, 52.
115. Ibid., 52–61.
116. Abana, interview with author.
117. Ibid.
118. Adjei, interview with author.
119. "2BN Off 3BN In," *Ghana Armed Forces News*, July 15, 1975, 8, GAF DPR.
120. Similar items described in Oduro-Kwarteng, *The Memoirs of a Colonel Retired*, 34; K. Oteng, "Memorandum from Lt. Col. K. Oteng to Jan G. Schumacher; March 7, 1977; Subject: Allegations against Ghanaian Troops" March 7, 1977, AG-035; S-1786 Office of the Force Commander—General Records 1973—1979, box S-0531-0070, folder S-1786-0009 Battallions and Units—Ghanbatt, UN ARMS.
121. Oteng, "Memorandum from Lt. Col. K. Oteng to Jan G. Schumacher; March 7, 1977; Subject: Allegations against Ghanaian Troops."

122. Oduro-Kwarteng, *The Memoirs of a Colonel Retired*, 61. Although it is tantalizing to imagine that the large Lebanese diaspora in Ghana facilitated these transactions, I found no evidence for it.
123. Ibid.
124. Abana, interview with author.
125. Chazan, *An Anatomy of Ghanaian Politics*, 6; Cooper, "The Dialectics of Decolonization," 407–414, 426; Killingray, "Imagined Martial Communities," 119; Baynham, "Introduction: Armed Forces in Africa," 7; Mann, *Native Sons*, 4.
126. Hutchful, "Institutional Decomposition," 251; Austin and Luckham, *Politicians and Soldiers in Ghana, 1966–1972*.
127. "Abongo Boy," "School Admissions," *Ghana Armed Forces Magazine*, October 1969, 3, FES.
128. Prince Osae-Djan, "Give Them Moral Support," *Ghana Armed Forces News*, February 15, 1975, 2.
129. In other places, Blay and his colleagues transcribed literally the stereotypical "Abongo" style of speech. Blay, "From Makana's Fanara Diary," 4; "Lt Cpl Abongo Is Back from No Man's Land," *Ghana Armed Forces News*, February 15, 1975, 2, GAF DPR; "Lt Cpl Abongo Gossips: Afro Man and Fine Fine 'Bobby,'" *Ghana Armed Forces News*, April 15, 1975, 2, GAF DPR.
130. Addae, *A Short History of Ghana Armed Forces*, 142; Addae, *History of Ghana Armed Forces*, 3:25.
131. The first female Ghanaian peacekeeper served in Lebanon in 1985. Owusu, "The Lone GHANBATT Lady," 5.
132. Cooper, "Industrial Man Goes to Africa," 130.
133. Ibid., 128.
134. Ibid., 131.
135. Parsons, *The African Rank-and-File*, 145–153. See also Echenberg, *Colonial Conscripts*, 121.
136. Comfort Akwagyiram, "Life of Army Wife with Hubby Away," *Ghana Armed Forces News*, September 15, 1976, 10, GAF DPR.
137. K. Adu-Bediaku, "The Wife in a Military World," *Ghana Armed Forces Magazine*, June 1968, 21, FES; Akwagyiram, "Life of Army Wife with Hubby Away," 10.
138. Gilman, *Mandarins of the Future*, 183.
139. See, for example, Mariama Bâ, *So Long a Letter*, African Writers Series 248 (London: Heinemann, 1989).
140. Adu-Bediaku, "The Wife in a Military World," 20.
141. Ibid.
142. Ibid., 20–21.
143. "Mrs. Soldier," *Ghana Armed Forces News*, September 15, 1975, 1.

144. Addae, *History of Ghana Armed Forces*, 3:26.
145. A remnant of the days when the British primarily employed Hausa men as the preferred "martial race" for the RWAFF. Ibid., 3:37; Killingray, "Imagined Martial Communities." See also Parsons, *The African Rank-and-File*, 53; Echenberg, *Colonial Conscripts*, 14; In *Ghana Armed Forces News*, the spelling is "magajia." Akwagyiram, "Life of Army Wife with Hubby Away," 10.
146. Addae, *History of Ghana Armed Forces*, 3:27, 38.
147. Ibid., 3:39.
148. Akwagyiram, "Life of Army Wife with Hubby Away," 10.
149. "2 BN in Action: The Women at Home and the Men on the Beat," *Ghana Armed Forces News*, December 15, 1975, 4, 8, GAF DPR.
150. "2 BN in Action."
151. Ibid., 4.
152. Akwagyiram, "Life of Army Wife with Hubby Away," 10.
153. Ibid.
154. Ibid.
155. Cooper, "Industrial Man Goes to Africa," 131; Dipesh Chakrabarty, "The Difference-Deferral of a Colonial Modernity," in *Tensions of Empire: Colonial Cultures in a Bourgeois World*, eds. Frederick Cooper and Ann Laura Stoler (Berkeley, CA: University of California Press, 1997), 378; Anna Davin, "Imperialism and Motherhood," in *Tensions of Empire: Colonial Cultures in a Bourgeois World*, eds. Frederick Cooper and Ann Laura Stoler (Berkeley, CA: University of California Press, 1997), 87–151.
156. Cooper, "Industrial Man Goes to Africa," 130; Parsons, *The African Rank-and-File*, 152; Mann, *Native Sons*, 178; Kevin K. Brown, "The Military and Social Change in Colonial Tanganyika: 1919–1964" (PhD Dissertation, Michigan State University, 2001), 150. Among "conscripted" African men in French armies, military service destabilized marriage patterns for different reasons. Echenberg, *Colonial Conscripts*, 84.
157. Cooper, "Industrial Man Goes to Africa," 131; Parsons, *The African Rank-and-File*, 146.
158. Parsons, *The African Rank-and-File*, 152.
159. Anyidoho, *My Journey . . . Every Step*, 52; Mann, *Native Sons*, 178; Oduro-Kwarteng, *The Memoirs of a Colonel Retired*, 17–18; Parsons, *The African Rank-and-File*, 152–162.
160. Oduro-Kwarteng, *The Memoirs of a Colonel Retired*, 17.
161. Ibid.
162. "Comds to Register All Wives," *Ghana Armed Forces News*, April 15, 1975, 8.
163. Cooper, *Colonialism in Question*, 131.

164. Ibid.
165. Cynthia H. Enloe, *Bananas, Beaches and Bases: Making Feminist Sense of International Politics* (Berkeley: University of California Press, 2000), chap. 9.
166. Pte. David Gamor, "Sentimental Journey," *Ghana Armed Forces News*, September 1, 1974, 4, GAF DPR.
167. Ibid.
168. Ibid.
169. On music and the memory of war, see Philip D. Beidler, *The Good War's Greatest Hits: World War II and American Remembering* (Athens: University of Georgia Press, 1998).
170. Katharine Gravino describes a similar process occurring among wives of American soldiers participating in UN peacekeeping operations in the early 1990s. Katharine Swift Gravino, David R. Segal, and Mady Wechsler Segal, "Lightfighters' Wives," in *Peacekeepers and Their Wives: American Participation in the Multinational Force and Observers*, eds. David R. Segal and Mady Wechsler Segal, Contributions in Military Studies, no. 147 (Westport, CT: Greenwood Press, 1993), 147, 155.
171. Hutchful, "Military Policy and Reform in Ghana," 253, 270.
172. Addae, *History of Ghana Armed Forces*, 3:91.
173. Anyidoho, *My Journey . . . Every Step*, 128.
174. Six hundred seventy-three students; six hundred ninety Ghanaian soldiers in Ghanbatt 19 (Fall 1982–Summer 1983). "Message from the Permanent Mission of Ghana to the UN Secretary General; August 9, 1983; Subject: Reimbursement for Specialists of GHANBATT" (New York, August 9, 1983), AG-20; S-0356 UNIFIL Subject Files, box 11, folder 7, UN ARMS.
175. United Nations Information Center, *Ghana: Four Decades*, 26–27.
176. Charles Moskos uses the term "supra nationalism" to indicate the specific version of internationalism UN soldiers in Cyprus reflected. Charles C. Moskos, *Peace Soldiers: The Sociology of a United Nations Military Force* (Chicago: University of Chicago Press, 1976), 106.
177. See, for example, Irwin, *Gordian Knot*; Ali A. Mazrui and Michael Tidy, *Nationalism and New States in Africa: From about 1935 to the Present* (London: Heinemann, 1984), 370–372. See also Westad, *The Global Cold War*; Borstelmann, *Apartheid's Reluctant Uncle*; Ninkovich, *The Wilsonian Century*, 287.
178. Nkrumah, *Africa Must Unite*, 194.
179. Quoted in Arkhurst, *Arms and African Development*, ix. For more on this conference, see Chapter 2.

180. Embassy of Ghana, "Ghana News, Vol. 14 No. 12 1985" (Washington, DC: Embassy of Ghana, December 1985), 1.
181. Ghana's ambassadors repeated this sentiment at monthly "troop contributors" meetings. See, for example, "Meeting with Countries Contributing Contingents to UNIFIL; October 12, 1982" October 12, 1982, AG-20; AG 020-001 Brian Urquhart, Under Secretary General of the UN for Special Political Affairs (1971–1985), box S-1066-0057, folder 1, UN ARMS; Special Committee On Peace-Keeping Operations, "Material On United Nations Observer Missions and Peace-Keeping Forces Authorized by the Security Council," 7.
182. Brian Urquhart, "Outgoing Code Cable from Urquhart to Siilasvuo; November 24, 1975; Subject: Additional Rifle Company" (New York, November 24, 1975), AG-31; S-0534 Administrative Files 1948–1987, box 467, folder S-1737-0171, UN ARMS; Special Committee On Peace-Keeping Operations, "Material On United Nations Observer Missions and Peace-Keeping Forces Authorized by the Security Council," 41; Kurt Waldheim, "Message from Kurt Waldheim (Sec-Gen) to Jomo Kenyatta (President of Kenya); March 18, 1974; Subject: Kenyan Contingent to UNEF II" March 18, 1976, AG-20, box S-1066-0006, folder 11, UN ARMS.
183. Kurt Waldheim, "Memorandum from Secretary-General to Permanent Representative of Ghana; November 1, 1977; Subject: Ghana Contingent to UNEF II" (New York, November 1, 1977), AG-20; S-0332 UNEF Subject Files, box S-0332-007, folder 6, UN ARMS.
184. "Message from Secretary General's Office to Permanent Representative of Ghana to the United Nations; 24 June 1981; Subject: Size of Ghana Contingent for UNIFIL" (New York, June 24, 1981), AG-20; S-0356 UNIFIL Subject Files, box 11, folder 6, UN ARMS; "Message from the Secretary-General to the Permanent Representative of Ghana; Subject: Increasing Size of Ghana Contingent; March 12, 1982."
185. "Meeting with Countries Contributing Contingents to UNIFIL; October 12, 1982."
186. Our Special Correspondent, "'Two Better' Takes On Tough Job," *Ghana Armed Forces News*, March 15, 1976, 4.
187. Ibid.
188. Capt. Victor Ansah, "The Suez Canal Story," *Ghana Armed Forces News*, May 15, 1975, 5, GAF DPR.
189. John S. K. Amissah, "UNIFIL- A Melting Pot," *Litani: UNIFIL News*, March 1985, 5, AG-20; S-1078 Personal Files, box S-1078-0089, folder 5—Litani—UNIFIL News Jan–Jun 1985, UN ARMS.
190. Ibid.

191. Ibid., 4–5.
192. Osofo is an alternate spelling of Asofo, the traditional Akan warrior class. Blay, "From Makana's Fanara Diary," 36.
193. Ibid.
194. "Glowing Tribute for Ghanbatt 5," *Ghana Armed Forces News*, September 15, 1976, 1.
195. "Ray of Hope from 'Sunrise,'" *Ghana Armed Forces News*, September 15, 1976, 10, GAF DPR.
196. Ibid.
197. Stephen Owusu, "A Praise-Worthy Performance," *Litani: UNIFIL News*, May 1985, 12–13, AG-20; S-1078 Personal Files, box S-1078-0089, folder 5—Litani—UNIFIL News Jan–Jun 1985, UN ARMS.
198. Ibid.
199. Ibid., 13. "Plaited," as it appears in the text.
200. Ibid., 12.
201. Ibid., 13.
202. Stephen Owusu, "United for Peace," *Litani: UNIFIL News*, March 1985, 21, AG-20; S-1078 Personal Files, box S-1078-0089, folder 5—Litani—UNIFIL News Jan–Jun 1985, UN ARMS.
203. Ibid.
204. "3BN On the Beat," *Ghana Armed Forces News*, April 15, 1975, 4, GAF DPR.
205. Capt. Victor Ansah, "Ghana Glitters Through the 'Barrels,'" *Ghana Armed Forces News*, September 1, 1974, 6, GAF DPR.
206. HOT BARRELS almost always appears in all caps, except after they departed the Sinai, when a reporter compared them to a sour drink.
207. Ansah, "Ghana Glitters Through the 'Barrels,'" 4.
208. Ibid. Style exactly as it appears in the text.
209. Ibid.
210. Ibid., 6.
211. "3BN On the Beat," 4.
212. Enninful, "To the War Zone and Back," 4–5.
213. Blay, "From Makana's Fanara Diary," 4.
214. A warrant officer was a senior noncommissioned officer, above most ranks of sergeant, but below the lowest rank of commissioned officer, lieutenant. Stephen Owusu, "Writing for Peace," *Litani: UNIFIL News*, May 1985, 6, AG-20; S-1078 Personal Files, box S-1078-0089, folder 5—Litani—UNIFIL News Jan-Jun 1985, UN ARMS; Stephen Owusu, *Writing for Peace* (Ghana: tecnic Channel, 2009).
215. Owusu, "Writing for Peace," May 1985, 2.
216. Ibid.

217. Ibid.
218. WOI Sowa Okpoti, "The Plea of the Peacekeeper," in *The Marakah*, vol. 2, *News of the Ghanaian Contingent* (Public Relations Detachment of GHANBATT 30, 1988), 15.
219. WOI Sowa Okpoti, "Mediterranean Sea," in *The Marakah*, vol. 2, *News of the Ghanaian Contingent* (Public Relations Detachment of GHANBATT 30, 1988), 15.
220. Pte. Ahwireng E. D. Atore, "Peace," *[Ghana] AFNEWS*, Last Quarter 1990, 12, GAF DPR.
221. Philip Cunliffe, *Legions of Peace: UN Peacekeepers from the Global South* (London: Hurst, 2013), 167–168.
222. Embassy of Ghana, "Erskine: Army Must Be Loyal to Govt.," 5.

## CHAPTER 4

1. *Senate Hearings Before the Committee on Appropriations; Foreign Operations, Export Financing, and Related Programs Appropriations, Fiscal Year 1991; H.R. 5114*, 101st Cong., 1st Sess. (June 22, 1990), 859, 909; General Accounting Office, *Security Assistance: Observations on the International Military Education and Training Program* (Washington, DC: GAO, National Security and International Affairs Division, June 14, 1990).
2. General Accounting Office, *Security Assistance*, 7.
3. Leahy himself had questioned Sec. of State Shultz extensively on American military assistance, including IMET, in 1986. *Foreign Assistance and Related Programs Appropriations for 1986. Hearings Before the Committee on Appropriations*. 99th Cong., 1st sess. (March 13, 1985), 122–138; *Hearing on H.R. 5114*, 909.
4. General Accounting Office, *Security Assistance*, 22.
5. Ibid., 23.
6. *Foreign Operations, Export Financing, and Related Programs Appropriations for 1990; Hearings Before a Subcommittee of the Committee on Appropriations; House of Representatives*, 101st Cong., 1st Sess. (March 15, 1989), pt. 5, 516. The report was published as a scholarly article in Spiro C. Manolas and Louis J. Samelson, "The United States International Military Education and Training (IMET) Program: A Report to Congress," *DISAM Journal of International Security Assistance Management* 12, no. 3 (Spring 1990): 12–47.
7. *Foreign Operations, Export Financing, and Related Programs Appropriations; Senate Hearings Before the Committee on Appropriations, Fiscal Year 1990*, Part 1, 101st Cong., 1st Sess. (April 4, 1989), pt. 1, 194. Emphasis in original; *Hearing on H.R. 5114*, 933.
8. *Foreign Operations, Export Financing, and Related Programs Appropriations for 1990, Hearing*, 516.

9. John A. Hannah, "New Directions in Foreign Aid for the 1970's," *American Journal of Agricultural Economics* 52, no. 2 (May 1, 1970): 303; Ninkovich, *The Wilsonian Century*, 234.
10. Rudolph A. Peterson, *U.S. Foreign Assistance in the 1970s: A New Approach; Report to the President from the Task Force on International Development* (Washington, DC: U.S. Govt. Print. Off., March 4, 1970), iv, 3, 6, 22.
11. Ibid., 18.
12. Ibid., 12.
13. Westad, *The Global Cold War*, 197; Guillermo A. O'Donnell, *Modernization and Bureaucratic-Authoritarianism: Studies in South American Politics*, Politics of Modernization Series (Berkeley: Institute of International Studies, University of California, 1979); Hyug Baeg Im, "The Rise of Bureaucratic Authoritarianism in South Korea," *World Politics* 39, no. 2 (January 1, 1987): 231–57.
14. See also, Mazower, *Governing the World*, chap. 11; A. Glenn Mower, *Human Rights and American Foreign Policy: The Carter and Reagan Experiences*, Studies in Human Rights, no. 7 (New York: Greenwood Press, 1987); Sarah B. Snyder, *Human Rights Activism and the End of the Cold War: A Transnational History of the Helsinki Network*, Human Rights in History (New York: Cambridge University Press, 2011).
15. Thomas Borstelmann, *The 1970s: A New Global History from Civil Rights to Economic Inequality* (Princeton, NJ: Princeton University Press, 2012), 204; Ninkovich, *The Wilsonian Century*, 234.
16. Lefever, "The Military Assistance Training Program," 361; Huntington, *Political Order in Changing Societies*, 193.
17. O'Donnell, *Modernization and Bureaucratic-Authoritarianism*, 99–103.
18. Bennett, "The Motivation for Military Intervention"; Bennett, "Epilogue: Malcontents in Uniform."
19. Westad, *The Global Cold War*, 197.
20. Gutteridge, "The Political Role of African Armed Forces," 103; Price, "Military Officers and Political Leadership"; Price, "A Theoretical Approach to Military Rule in New States."
21. Ernest W. Lefever, "Arms Transfers, Military Training, and Domestic Politics," in *Arms Transfers in the Modern World*, eds. Stephanie G. Neuman and Robert E. Harkavy (New York: Praeger, 1979), 279–291; Huntington, *The Soldier and the State*, chap. 4; Mazrui, "Soldiers as Traditionalizers"; A. G. G. Gingyera-Pincywa and Ali A. Mazrui, "Regional Development and Regional Disarmament: Some African Perspectives," in *Arms and African Development: Proceedings of the First Pan-African Citizens' Conference*, ed.

Frederick S. Arkhurst, Praeger Special Studies in International Economics and Development (New York: Praeger, 1972), 32.

22. DSAA, *Congressional Presentation, Security Assistance Programs*, FY 1980 (1979), 16.
23. Donald M. Fraser, "Human Rights and U.S. Foreign Policy: Some Basic Questions Regarding Principles and Practice," *International Studies Quarterly* 23, no. 2 (June 1, 1979): 174–85; Snyder, *Human Rights Activism and the End of the Cold War*, 9; Peterson, *U.S. Foreign Assistance in the 1970s*; DSAA, *Congressional Presentation, Security Assistance Programs*, FY 1980; Jimmy Carter, "Presidential Directive NSC-30; Subject: Human Rights," February 17, 1978, Federation of American Scientists Intelligence Resource Program, *www.fas.org/irp/offdocs/pd/index.html*.
24. "Department of State Research Study; Subject: Proceedings of Conference on Security Assistance; December 1, 1972" (Report, Washington, DC, December 1, 1972), 7, RG 59 SNF 1970–1973; Political and Defense, box 1848, folder DEF 19 1/1/70, NACP.
25. Ibid.
26. Huntington, *Political Order in Changing Societies*, 193.
27. Lefever, "The Military Assistance Training Program," 361.
28. G. Warren Nutter, "Memorandum from G. Warren Nutter, Assistant Sec. of Defense for International Security Affairs, to the Chairman of Joint Chiefs of Staff, July 7, 1971; Subject: Training of Foreign Military Personnel Under the Security Assistance Program" (Memorandum, July 7, 1971), RG 330, accession no. UD/WW 158, box 2, folder 35, "Review of Military Assistance Training, Informational Program and Orientation Tours, 23 December 1971," NACP.
29. Neil H. Campbell and Elliot B. Strauss, "Memorandum from Neil H. Campbell and Elliot B. Strauss, Office of the Inspector General of Foreign Assistance, U.S. Department of State, to Anthony Faunce, Acting Inspector General, Department of State, November 15, 1971, Subject: Review of Military Assistance Training—Informational Program and Orientation Tours" (November 15, 1971), RG 330, accession no. UD/WW 158, box 2, folder 35, "Review of Military Assistance Training, Informational Program and Orientation Tours, 23 December 1971," NACP.
30. Ibid., 2.
31. Ibid., 3.
32. George M. Seignious, II, "Memorandum from George M. Seignious, Director Defense Security Assistance Agency to Hon. Anthony Faunce, Acting IG, Foreign Assistance, Dept. of State; April 11, 1972; Subject: Report on Review of Military Assistance Training—Informational Program

and Orientation Tours" (April 11, 1972), 3, RG 330, accession no. UD/WW 158, box 2, folder 35, "Review of Military Assistance Training, Informational Program and Orientation Tours, 23 December 1971," NACP.

33. Lillian Craig Harris, *Libya: Qadhafi's Revolution and the Modern State* (Boulder, CO: Westview Press, 1986), 47; David Blundy and Andrew Lycett, *Qaddafi and the Libyan Revolution* (Boston: Little, Brown, 1987), 49.
34. DSAA, *Congressional Presentation, Security Assistance Programs*, FY 1981, 7, 503; Campbell and Strauss, "Office of the IG for Foreign Assistance Memo, Nov. 15, 1971, Subj: Review of Mil. Asst. Training—Informational Program," 4.
35. Campbell and Strauss, "Office of the IG for Foreign Assistance Memo, Nov. 15, 1971, Subj: Review of Mil. Asst. Training—Informational Program," cover letter.
36. Achel, interview with author. For more on sponsor families in the Kansas City area, see Clune, "The Abongo Abroad" (PhD Dissertation, University of Kansas, 2014), ch. 3.
37. Campbell and Strauss, "Office of the IG for Foreign Assistance Memo, Nov. 15, 1971, Subj: Review of Mil. Asst. Training—Informational Program," 7.
38. Iriye, "Internationalizing International History," 53.
39. Campbell and Strauss, "Office of the IG for Foreign Assistance Memo, Nov. 15, 1971, Subj: Review of Mil. Asst. Training—Informational Program," 2.
40. Ibid., 13, 7. No such screening occurred for accompanying spouses. Local communities often filled that need.
41. "Unclassified Message from USCINCSO (Southern Command) to Secretary of Defense, 3 January 1972, Subject: 'Dept. of State Inspector General Foreign Assistance (IFA 430945, 'Review of Military Assistance Training—Information Program and Orientation Tours''" (January 3, 1972), RG 330, accession no. UD/WW 158, box 2, folder 35, "Review of Military Assistance Training, Informational Program and Orientation Tours, 23 December 1971," NACP.
42. Ibid.
43. Ibid.
44. Ibid.
45. Nutter, "Nutter to CJCS; July 7, 1971; Subj: Training Foreign Military Personnel."
46. Immanuel Maurice Wallerstein, "Modernization: Requiescat in Pace," in *The Capitalist World-Economy: Essays*, Studies in Modern Capitalism (Cambridge: Cambridge University Press, 1979), 132–37; Latham, *Modernization as Ideology*, 8.

47. Atlee R. Ellis, "Memorandum from Atlee R. Ellis, U.S. Air Force, Deputy Directory of Military Assistance and Sales to Defense Security Assistance Agency (Comptroller), 23 December 1971, Subject: IGA Report, 'Review of Military Assistance Training—Informational Program and Orientation Tours'" December 23, 1971, RG 330, accession no. UD/WW 158, box 2, folder 35, "Review of Military Assistance Training, Informational Program and Orientation Tours, 23 December 1971," NACP.
48. Ibid.
49. Peterson, *U.S. Foreign Assistance in the 1970s*, 3.
50. Amos et al., *US Training of Foreign Military Personnel, Volume 2*, para. 2.2. Policy makers used the phrase "international military education and training" for over a decade, but it became an official title in 1976. I use the acronym IMET when describing the specific DoD/State-sponsored program, not the concept of international training.
51. Manolas and Samelson, "The IMET Program: Report to Congress," 32.
52. Ibid.
53. Lefever, "The Military Assistance Training Program," 87.
54. Amos et al., *U.S. Training of Foreign Military Personnel, Volume 2*, 24. On how the School of the Americas straddles the line between overseas "military" training and US-based "professional" training, see Gill, *The School of the Americas*.
55. Lefever, "The Military Assistance Training Program," 87.
56. Amos et al., *U.S. Training of Foreign Military Personnel, Volume 2*, 23–26.
57. Ibid., 8.
58. *International Security Assistance and Arms Export Control Act of 1976. Hearings Before the Committee on International Relations*, 94th Cong., 2d Sess. (March 23–30, 1976), 126.
59. Ibid.
60. Black, "Black to Kissinger; Mar. 12, 1976; Subj: Changes in Pricing of Military Training."
61. *International Security Assistance and Arms Export Control Act of 1976. Hearings Before the Committee on International Relations,* 4.
62. Ibid., 23.
63. Ibid.
64. Richard F. Grimmett and Mark P. Sullivan, *U.S. Army School of the Americas: Background and Congressional Concerns,* CRS Report for Congress (Washington, DC: Congressional Research Service, 2001), 4.
65. Timothy Niblock, "A New Political System in Sudan," *African Affairs* 73, no. 293 (October 1974): 408; "REAL People to People. . . . Hosts and

Exchange Visits—JUST A SAMPLING OF PTP ACTIVITIVES," *Contact: Greater Kansas City People-to-People Council Newsletter*, May 1979, binder: Contact 1970s, GKCPTP; Achel, interview with author.

66. Gaafar M. Nimeiri, "Letter from Gaafar Nimeiri to Helen and Bob Black, February 24, 1970" (Khartoum, Sudan, February 24, 1970), Helen Black Collection, box 1, folder 3 Sudan-Nimeiri-1970, CARL; "Sudan's Chief: Well-Meaning, Awkward," *Washington Post*, June 20, 1971, A24, Helen Black Collection, box 1, folder 3 Sudan-Nimeiri-1970, CARL; Warren E. Hearnes, "Letter from Warren E. Hearns, Governor, to President Gaafar Mohamed Nimeri [sic], May 12, 1970" (Jefferson City, MO, May 12, 1970), Helen Black Collection, box 1, folder 3 Sudan-Nimeiri-1970, CARL.
67. Robert Pearman, "Ft. Leavenworth 'Alums' in the News," *Kansas City Times*, January 24, 1972, Helen Black Collection, box 27, folder 4—1972, CARL.
68. "Helen Black," *Contact: Greater Kansas City People-to-People Council Newsletter*, October 1978, 2, binder: Contact 1970s, GKCPTP.
69. Jesse Dillon Savage and Jonathan D. Caverley, "Foreign Military Training and Coup Propensity" (unpublished article, October 1, 2015), 27, *www.jonathan-caverley.com/uploads/2/9/7/2/29726853/caverleysavage_sept17_2015.pdf*.
70. *International Security Assistance and Arms Export Control Act of 1976. Hearings Before the Committee on International Relations*, 38.
71. Ibid.
72. *International Security Assistance Programs: Hearings Before the Subcommittee on Foreign Assistance of the Committee on Foreign Relations, United States Senate, Ninety-Fifth Congress, Second Session on S. 2846 To Amend the Foreign Assistance Act of 1961 and the Arms Export Control Act, and for Other Purposes*, 95th Cong., 2d Sess. (April 25–26; May 1–2, 1978), 45.
73. Ibid.
74. Ibid., 12.
75. Daniel P. Moynihan, "The United States in Opposition," *Commentary* 59, no. 3 (March 1975): 41.
76. Ibid., 42; Mazower, *Governing the World*, 309.
77. Snyder, *Human Rights Activism and the End of the Cold War*, 9.
78. Ninkovich, *The Wilsonian Century*, 234.
79. John N. Irwin II, "Secret Memorandum (Draft) For the President from the NSC Undersecretaries Committee and Undersecretary of State John N. Irwin II Subject: Study of Security Assistance (NSDM 76)" (Memorandum, Washington, DC, December 5, 1970), RG 59 SNF 1970–1973, Political and Defense, box 1848, folder DEF 19 1/1/70, NACP; "Department of State Research Study; Subject: Proceedings of Conference On Security Assistance; December 1, 1972."

80. *Public Law 95-384, 95 Congress, Session 2, an Act: To Amend the Foreign Assistance Act of 1961 and the Arms Export Control Act to Authorize International Security Assistance Programs for Fiscal Year 1979, and for Other Purposes. Short Title: International Security Assistance Act of 1978, 92 Stat,* 1978, 732, 736.
81. US Department of Defense, "DoDD 5410.17, 1 March 1985."
82. Amos et al., *U.S. Training of Foreign Military Personnel, Volume 2*, 86.
83. Ibid., i.
84. Ibid., ii.
85. DSAA, *Congressional Presentation, Security Assistance Programs*, FY 1980, 16.
86. Amos et al., *US Training of Foreign Military Personnel, Volume 2*, ii.
87. See, for example, US Department of Defense, "Department of Defense Directive 5410.17 'An Informational Program for Foreign Military Trainees and Visitors in the United States,'" January 15, 1965, Combined Arms Research Library; Benson in *Hearings Before the Subcommittee on Foreign Assistance of the Committee on Foreign Relations, on S. 2846*.
88. Black, "Black to Kissinger; Mar. 12, 1976; Subj: Changes in Pricing of Military Training."
89. Amos et al., *U.S. Training of Foreign Military Personnel, Volume 2*, 69.
90. Ibid., viii.
91. Ibid., 73.
92. Ibid., 72. Emphasis is mine.
93. Ibid.
94. Ibid., viii.
95. Some notable examples include Holden, "The Real Diplomacy of Violence"; Henry S. Bienen, "Armed Forces and National Modernization: Continuing the Debate," *Comparative Politics* 16, no. 1 (October 1, 1983): 1–16; Fitch, "The Political Impact of U.S. Military Aid to Latin America"; Westad, *The Global Cold War*; Arlinghaus, *Military Development in Africa*; Kolko, *Confronting the Third World*; Chomsky, *American Power and the New Mandarins*; Joseph P. Smaldone, "U.S. Arms Transfers and Security Assistance Programs in Africa: A Review and Policy Perspective," in *Arms for Africa: Military Assistance and Foreign Policy in the Developing World*, ed. Bruce E. Arlinghaus (Lexington, MA: D. C. Heath, 1983), 179–220; Robert M. Rosh, "Third World Militarization: Security Webs and the States They Ensnare," *Journal of Conflict Resolution* 32, no. 4 (December 1, 1988): 671–98; Wiley, "Militarizing Africa and African Studies and the U.S. Africanist Response"; Bienen, *The Military and Modernization*; Tom Wicker, "A Theological War," *Diplomatic History* 20, no. 3 (1996): 445–49.

96. Amos et al., *U.S. Training of Foreign Military Personnel, Volume 2*, 12.
97. Ibid., 13.
98. Ibid., 14.
99. Ibid., iii.
100. Ibid., iv. Other foreign military personnel purchased tuition at American military schools as part of the Foreign Military Sales program, which was not considered military assistance and followed different legislative rules (even if the individuals' experiences at American schools and bases were similar.)
101. Martha Finnemore, "Redefining Development at the World Bank," in *International Development and the Social Sciences: Essays on the History and Politics of Knowledge*, eds. Frederick Cooper and Randall M. Packard (Berkeley: University of California Press, 1997), 220.
102. Amos et al., *U.S. Training of Foreign Military Personnel, Volume 2*, 15.
103. Ibid., 23.
104. Kissinger, "Confidential Telegram from Secretary of State to All American Diplomatic Posts, 3 March 1976, Subject: Changes in Foreign Military Training Program"; Black, "Black to Kissinger; Mar. 12, 1976; Subj: Changes in Pricing of Military Training."
105. Amos et al., *U.S. Training of Foreign Military Personnel, Volume 2*, 16.
106. Ibid., 74.
107. Ibid., iii. Emphasis is mine.
108. Ibid., 74.
109. Lefever, "Arms Transfers, Military Training, and Domestic Politics," 279; Lefever, "The Military Assistance Training Program."
110. Lefever, "Arms Transfers, Military Training, and Domestic Politics," 291.
111. See Seigel, *Uneven Encounters*, xii.
112. DSAA, *Congressional Presentation, Security Assistance Programs*, FY 1981, 1, 16; DSAA, *Congressional Presentation, Security Assistance Programs*, FY 1988 (1987), 18.
113. On the "boomerang effect," see Daniel C. Thomas, "Boomerangs and Superpowers: International Norms, Transnational Networks and US Foreign Policy," *Cambridge Review of International Affairs* 15, no. 1 (April 2002): 26; Snyder, *Human Rights Activism and the End of the Cold War*, 11.
114. Data from congressional presentations, fiscal years 1978 through 1989. At most, the United States spent $281,000 (1982) on forty-three total students; the lowest: $188,000 (1979) on twenty-three total students.
115. DSAA, *Congressional Presentation, Security Assistance Programs*, FY 1989 (1988), 163.
116. DSAA, *Congressional Presentation, Security Assistance Programs*, FY 1982 (1981), 295.

117. Total economic aid for Ghana taken from congressional presentations, fiscal years 1980–81 through 1986–87. These reported the "actual" amount spent for the prior year, the current year's "expected" budget, and the projected budget for the fiscal year of the document's title. Zero dollars in economic aid went to Ghana in 1980, which I included in the average. Otherwise, the average would be about US$17 million per year.
118. DSAA, *Congressional Presentation, Security Assistance Programs*, FY 1980, 279.
119. Ibid.
120. Ibid.
121. Released in 1980, this was: DSAA, *Congressional Presentation, Security Assistance Programs*, FY 1981.
122. There is no US military equivalent for the rank of flight lieutenant, but it is something close to a junior captain in the US Army, although Rawlings had been a commissioned officer in the Ghana Air Force for ten years prior to 1979. In the US Army or US Air Force, normally an officer would have been promoted from captain to major with about that same tenure.
123. DSAA, *Congressional Presentation, Security Assistance Programs*, FY 1981, 277.
124. Ibid.
125. Ibid.
126. DSAA, *Congressional Presentation, Security Assistance Programs*, FY 1982, 295.
127. Emmanuel Hansen and Paul Collins, "The Army, the State, and the 'Rawlings Revolution' in Ghana," *African Affairs* 79, no. 314 (January 1, 1980): 4; See Austin, "The Ghana Armed Forces and Ghanaian Society," 101; Baynham, "Divide et Impera," 634; Decalo, "Military Rule in Africa: Etiology and Morphology," 56; Emmanuel Hansen, *Ghana Under Rawlings: Early Years* (Lagos, Nigeria: Malthouse Press, 1991); Hutchful, "Institutional Decomposition," 237.
128. Embassy of Ghana, "Rawlings Explains Motive for Military Take-Over," *Ghana News*, January 1982, 6; Embassy of Ghana, "Holy War Still On — Rawlings," *Ghana News*, January 1983.
129. Embassy of Ghana, "Rawlings Explains Motive for Military Take-Over," 6–7.
130. Embassy of Ghana, "Days of Elitist Officers Are No More," *Ghana News*, June 1982, 4.
131. Embassy of Ghana, "Relations with Libya Restored," *Ghana News*, January 1982, 9; Embassy of Ghana, "Bright Future for Ghana—Soviet Relations," *Ghana News*, May 1984; Embassy of Ghana, "Soviets Declare Support," *Ghana News*, January 1982, 9; Embassy of Ghana, "Iranians Hold Talks on Trade," *Ghana News*, June 1982, 4; Embassy of Ghana, "Cultural Exchange Pact Signed," *Ghana News*, January 1983, 2.
132. Hutchful, "Institutional Decomposition," 248–251.

133. Ibid., 212.
134. DSAA, *Congressional Presentation, Security Assistance Programs*, FY 1983 (1982), 305.
135. DSAA, *Congressional Presentation, Security Assistance Programs*, FY 1980, 16.
136. DSAA, *Congressional Presentation, Security Assistance Programs*, FY 1985 (1984) 27.
137. DSAA, *Congressional Presentation, Security Assistance Programs*, FY 1980, 16. Notice here the congressional presentation repeated the CRS's 1979 report verbatim, adding the phrase "essentially cost free." It is not clear who was copying whom.
138. *Security and Development Assistance: Hearings before the Committee on Foreign Relations, US Senate*, 98th Cong., 1st Sess. (March 8, 1983), 426.
139. DSAA, *Congressional Presentation, Security Assistance Programs*, FY 1984 (1983), 242.
140. DSAA, *Congressional Presentation, Security Assistance Programs*, FY 1985, 252.
141. DSAA, *Congressional Presentation, Security Assistance Programs*, FY 1986 (1985), 270.
142. DSAA, *Congressional Presentation, Security Assistance Programs*, FY 1987, Volume 2 (1986), 178.
143. Hansen and Collins, "The Army, the State, and the 'Rawlings Revolution' in Ghana"; Embassy of Ghana, "Rawlings Explains Motive for Military Take-Over"; E. Gyimah-Boadi and Donald Rothchild, "Rawlings, Populism, and the Civil Liberties Tradition in Ghana," *Issue: A Journal of Opinion* 12, no. 3/4 (October 1, 1982): 64–69.
144. See, for example, Cullather, *The Hungry World*, 5.
145. Adekson, "Pay, Promotion, and Other Self-Interests of Military Intervention in Politics."
146. Mazower, *Governing the World*, 346; Westad, *The Global Cold War*, 331.
147. Hunt, *Ideology and U.S. Foreign Policy*, 187; Westad, *The Global Cold War*, 333–334; McCormick, *America's Half-Century*, 226.
148. Military Assistance Program (grants of weapons, equipment, etc. not including training): $110 million in 1980, $805 million in 1985; Economic Support Fund (economic assistance to strategically important states): $1.94 billion in 1980, $4 billion in 1985. The total ($9 billion) includes loans and loan guarantees for the massive ($3 billion to $5.7 billion) Foreign Military Sales programs, some of which would theoretically be repaid. DSAA, *Congressional Presentation, Security Assistance Programs*, FY 1989, 386–88; In 1985, roughly 45 percent of all American military and economic assistance went to Egypt and Israel. DSAA, *Congressional Presentation, Security Assistance Programs*, FY 1985.

149. DSAA, *Congressional Presentation, Security Assistance Programs*, FY 1989, 388.
150. Westad, *The Global Cold War*, 359–361.
151. DSAA, *Congressional Presentation, Security Assistance Programs*, FY 1980, 279.
152. McCormick, *America's Half-Century*, 235; Westad, *The Global Cold War*, 359–360.
153. See, for example, Westad, *The Global Cold War*, chap. 9. As a counterpoint, Sen. Helms (R-NC) attacked US military aid on different principles, criticizing US intervention in El Salvador for being *too* socialist and criticizing the US emphasis on human rights. *International Security and Development Cooperation Act of 1984; Report of the Committee on Foreign Relations, United States Senate on S. 2582, to Provide a Supplemental Authorization of Appropriations for the Fiscal Year 1984 for Certain Foreign Assistance Programs; to Amend the Foreign Assistance Act of 1961, the Arms Export Control Act, and Other Acts to Authorize Appropriations for the Fiscal Year 1985 for International Security and Development Assistance, the Peace Corps and the International Development Association; and for Other Purposes*, 98th Cong., 2d Sess. (April 18, 1984), 70.
154. See Ruttan, *United States Development Assistance Policy.*
155. *Foreign Assistance and Related Programs Appropriations for 1986*, 99th Cong., 1st Sess., (March 26, 1985), 517.
156. Ibid.
157. Ibid., 517–518.
158. DSAA, *Congressional Presentation, Security Assistance Programs*, FY 1985, 5.
159. Sec. of State Shultz, Lt. Gen. Gast (Defense) and Under Sec. Schneider field mostly questions on human rights in *Foreign Assistance and Related Programs Appropriations for 1986*, 128–138, 169–280.
160. In 1985, for example, 40 percent of the $9 billion total US Security Assistance budget went to Israel and Egypt. DSAA, *Congressional Presentation, Security Assistance Programs*, FY 1985, 13.
161. In the late 1960s IMET grew to more than 10 percent of total military assistance funds; in the late 1970s, from 1978 to 1981, it was more than 25 percent. DSAA, *Congressional Presentation, Security Assistance Programs*, FY 1989, 388.
162. Foreign Operations, Export Financing, and Related Programs Appropriation Bill, 1990, Report to Accompany H.R. 2939. S. Rep. No. 101-131, at 162 (1990).
163. Ibid.
164. Ibid.
165. General Accounting Office, *Security Assistance, Hearing on H.R. 5114*, 908.
166. General Accounting Office, *Security Assistance*, 2–4.

167. Ibid., 22.
168. Manolas and Samelson, "The IMET Program: Report to Congress," 1. Emphasis in original.
169. Ibid.
170. Ibid., 20.
171. Conference Report on International Security Assistance and Arms Export Control Act of 1976, H.R. 13680 to Amend the Foreign Assistance Act of 1961 and the Foreign Military Sales Act, and for Other Purposes, H.R. Rep. 94-1272, at 50–51; *International Security Assistance and Arms Export Control Act of 1976. Hearings Before the Committee on International Relations,* 23; The International Security Assistance and Arms Export Control Act of 1977. Report from the Committee on Foreign Relations, 95th Cong., 1st Sess., May 16, 1977, 7; House Committee on Foreign Affairs, Foreign Assistance Act of 1969; Report of the Committee on Foreign Affairs Together with Minority, Supplemental and Additional Views on H.R. 14580 to Promote the Foreign Policy, Security, and General Welfare of the United States by Assisting Peoples of the World, Etc., H.R. Rep. 91-611, at 10; *Senate Committee on Foreign Relations (95th Cong.) Hearings to Amend the FAA of 1961*, 86; International Security and Development Cooperation Act of 1984: Report of the Committee of Foreign Affairs, Together with Additional and Supplemental View [including Cost Estimate of the Congressional Budget Office] on H.R. 5119, to Authorize International Development and Security Assistance Programs and Peace Corps Programs for Fiscal Year 1985, and for Other Purposes, H.R. Rep. 98-628, at 20.
172. Manolas and Samelson, "The IMET Program: Report to Congress," 5.
173. *Hearing on H.R. 5114,* 891.
174. Ibid., 892.
175. Gilman, *Mandarins of the Future*, x.
176. Cullather, *The Hungry World*, 132.
177. US President's Committee to Study the US Military Assistance Program (1958–1959), "Draper Committee Final Report," 43.
178. Amos et al., *U.S. Training of Foreign Military Personnel, Volume 2,* 2–14.

## CONCLUSION

1. DSAA, *Congressional Presentation, Security Assistance Programs*, FY 1981, 510; Manolas and Samelson, "The IMET Program: Report to Congress," 35.
2. Jennifer Morrison Taw, *The Effectiveness of Training International Military Students in Internal Defense and Development,* executive summary (Santa Monica, CA: RAND, 1993), iii.
3. Jennifer Morrison Taw and William H. McCoy, *International Military Student Training: Beyond Tactics*, a RAND note (Santa Monica, CA: RAND, 1993), 11.

4. Taw, *Effectiveness of Training International Military Students,* executive summary, 18.
5. McCoy, *Senegal and Liberia*, 5.
6. Between 1976 and 1989, 413 Liberians and 451 Ghanaians. Manolas and Samelson, "The IMET Program: Report to Congress," 35; McCoy, *Senegal and Liberia*, 25, 61.
7. Savage and Caverley, "Foreign Military Training and Coup Propensity," 27.
8. From 1976 to 1989, 201 Senegalese. McCoy, *Senegal and Liberia*, 8; Manolas and Samelson, "The IMET Program: Report to Congress," 35.
9. Nelson-Pallmeyer, *School of Assassins*; Gill, *The School of the Americas*; Jennifer Morrison Taw, *Thailand and the Philippines: Case Studies in U.S. IMET Training and Its Role in Internal Defense and Development* (Santa Monica, CA: RAND, 1994), xiv; Taw, *Effectiveness of Training International Military Students, Executive Summary*, vii; McCoy repeats "a rumor," spread via "a human rights newsletter," that Marcos's top torturer (Lt. Col. Rolando Abadilla) studied at CGSC. McCoy, *Closer than Brothers*, 190; Holden, "The Real Diplomacy of Violence"; Bartholomew, *Senate Committee on Appropriations: Foreign Operations for FY 1991*, 891.
10. Weschler, "A Reporter at Large: The Great Exception I—Liberty," 48.
11. Ibid., 46.
12. Iriye, *Cultural Internationalism and World Order*, 2.

# Bibliography


## ARCHIVAL SOURCES AND COLLECTIONS:

Bibliothek der Friedrich-Ebert-Stiftung, Bonn, Germany (FES)
City of Leavenworth Public Library *Leavenworth Times* Microfilm Collection
Directorate of Public Relations, Ghana Armed Forces, Burma Camp, Accra (DPR)
Fort Leavenworth Combined Arms Research Library (CARL)
International Officers Collection
Helen Black Collection
George Padmore Research Library, Accra
Ghana Armed Forces Command and Staff College Library, Teshie (GAFCSC)
*Ghana News*, Publication of the Ghana Embassy in the United States
Ghana Public Records and Archives Administration (PRAAD)
Greater Kansas City Chapter of People-to-People International, Kansas City, MO (GKCPTP)
United Nations Archives and Records Management Section (ARMS)
AG-020 United Nations Office for Special Political Affairs 1946–1991
AG-035 United Nations—Emergency Force II (UNEF) 1973–1979
AG-031 Field Service and Successor Bodies (1948–1982)
AG-006 Secretary-General Kurt Waldheim (1972–1981)
United States National Archives at College Park, MD (NACP)
Record Group 59 General Records of the Department of State Central Decimal Files (CDF), Central Foreign Policy Files (CFPF), Subject Numeric Files (SNF)
Record Group 330 Records of the Office of the Secretary of Defense (RG 330)

## INTERVIEWS AND ORAL HISTORIES

Michael Abana* (Colonel, retired), former Faculty, Ghana Armed Forces Staff College, July 18, 2012.

Ignatius Achel* (Brigadier, retired), former Commandant, Ghana Armed Forces Staff College, July 19, 2012.

Lawrence Adjei* (Lieutenant General, retired), former Chief of Defense Staff, Ghana Armed Forces, July 20, 2012.

Kwabena Bamfo* (Colonel, Ghana Army), Director of Public Relations, Ghana Armed Forces, July 18, 2012.

Christian Kumi* (Wing Commander—equivalent to Lieutenant Colonel, Ghana Air Force), Kofi Annan International Peacekeeping Training Center, July 16, 2012.

Tom Gbologah* (Colonel, Ghana Army), Kofi Annan International Peacekeeping Training Center, July 16, 2012.

*Names have been changed to ensure anonymity in accordance with University of Kansas Human Subjects Committee recommendations. Interview templates approved in case numbers: HSCL #19993 (April 9, 2012) and STUDY00000444 (December 3, 2013).

## US GOVERNMENT DOCUMENTS AND REPORTS

### *FOREIGN RELATIONS OF THE UNITED STATES SERIES (FRUS)*

Claussen, Paul, Evan M. Duncan, and Jeffrey A. Soukup, eds. *Organization of Foreign Policy; Information Policy; United Nations; Scientific Matters*. Vol. XXV. Foreign Relations of the United States, 1961–1963. Washington, DC: US Government Printing Office, 2001.

Coffman, Suzanne E., Edward C. Keefer, and Harriet Dashiell Schwar, eds. *Foreign Economic Policy*. Vol. IV. Foreign Relations of the United States, 1958–1960. Washington, DC: US Government Printing Office, 1991.

Gerakas, Evans, David S Patterson, William F. Sanford, Jr., and Carolyn B. Yee, eds. *Foreign Economic Policy*. Vol. IX. Foreign Relations of the United States, 1961–1963. Washington, DC: US Government Printing Office, 1995.

Howland, Nina Davis, ed. *Africa*. Vol. XXI. Foreign Relations of the United States, 1961–1963. Washington, DC: US Government Printing Office, 1995.

———, ed. *Africa*. Vol. XXIV. Foreign Relations of the United States, 1964–1968. Washington, DC: US Government Printing Office, 1995.

### *CONGRESSIONAL HEARINGS*

US Congress. House. *Foreign Assistance and Related Programs Appropriations for 1986. Hearings Before the Committee on Appropriations*. 99th Cong., 1st Sess. (March 26, 1985).

US Congress. House. *Foreign Operations, Export Financing, and Related Programs Appropriations for 1990; Hearings Before a Subcommittee of the Committee on Appropriations; Part 5*. 101st Cong., 1st Sess. (March 15, 1989).

US Congress. House. *International Security Assistance and Arms Export Control Act of 1976. Hearings Before the Committee on International Relations*. 94th Cong., 2d Sess. (March 23–31, 1976).

US Congress. Senate. *Foreign Assistance and Related Programs Appropriations for 1986. Hearings Before the Committee on Appropriations*. 99th Cong., 1st Sess. (March 13, 1985).

US Congress. Senate. *Foreign Operations, Export Financing, and Related Programs Appropriations; Hearings Before the Committee on Appropriations, Fiscal Year 1990; Part 1*. 101st Cong., 1st Sess. (April 4, 1989).

US Congress. Senate. *Hearings Before the Committee on Appropriations; Foreign Operations, Export Financing, and Related Programs Appropriations, Fiscal Year 1991; H. R. 5114*. 101st Cong., 1st Sess. (June 22, 1990).

US Congress. Senate. *International Development and Security. Hearings Before the Committee on Foreign Relations, United States Senate. 87th Congress, First Sess. on S. 1983. A Bill to Promote the Foreign Policy, Security, and General Welfare of the United States by Assisting Peoples of the World in Their Efforts Towards Economic and Social Development and Internal and External Security, and for Other Purposes*. 2 vols. 87th Cong., 1st Sess. (May 31, June 5–9; 12–16; 19, 20, and 27, 1961).

US Congress. Senate. *International Security Assistance Programs: Hearings Before the Subcommittee on Foreign Assistance of the Committee on Foreign Relations, on S. 2846 to Amend the Foreign Assistance Act of 1961 and the Arms Export Control Act, and for Other Purposes*. 95th Cong., 2d Sess. (April 25–26; May 1–2, 1978).

US Congress. Senate. *Security and Development Assistance: Hearings Before the Committee on Foreign Relations*. 98th Cong., 1st Sess. (March 8, 1983).

### CONGRESSIONAL REPORTS

US Congress. House. *Conference Report on International Security Assistance and Arms Export Control Act of 1976, H.R. 13680 to Amend the Foreign Assistance Act of 1961 and the Foreign Military Sales Act, and for Other Purposes*, H.R. Rep. No. 94-1272 (1976).

US Congress. House. *Foreign Assistance Act of 1969; Report of the Committee on Foreign Affairs Together with Minority, Supplemental and Additional Views on H.R. 14580 to Promote the Foreign Policy, Security, and General Welfare of the United States by Assisting Peoples of the World, Etc.*, H.R. Rep. No. 91-611 (1969).

US Congress. House. *International Security and Development Cooperation Act of 1984; Report of the Committee of Foreign Affairs, Together with Additional and Supplemental View [Including Cost Estimate of the Congressional Budget Office] on H.R. 5119, to Authorize International Development and Security Assistance Programs and Peace Corps Programs for Fiscal Year 1985, and for Other Purposes*, H.R. Rep. No. 98-628 (1984).

US Congress. House. *Report of the Committee on Foreign Affairs on H.R. 11510 to Amend Further the Mutual Security Act of 1954*, H.R. Rep. No. 86-1464 (1960).

US Congress. House. *Report of the Special Study Mission to Asia, Western Pacific, Middle East, Southern Europe, and North Africa*, H.R. Rep. No. 86-1386 (1960).

US Congress. House. *Use of Defense Support Funds for Economic and Political Purposes. Twenty-First Report by the Committee on Government Operations*, H.R. Rep. No. 85-1374 (1958).

US Congress. Senate. *A Report on United States Foreign Operations by Honorable Allen J. Ellender, United States Senator from the State of Louisiana*, S. Rep. No. 87-20 (1961).

US Congress. Senate. *Foreign Assistance Act of 1961; Report of the Committee on Foreign Relations, United States Senate, on S. 1983*, S. Rep. No. 87-612 (1961).

US Congress. Senate. *Foreign Operations, Export Financing, and Related Programs Appropriation Bill, 1990, Report to Accompany H.R. 2939*, S. Rep. No. 101-131 (1989).

US Congress. Senate. *International Security and Development Cooperation Act of 1984; Report of the Committee on Foreign Relations, United States Senate on S. 2582, to Provide a Supplemental Authorization of Appropriations for the Fiscal Year 1984 for Certain Foreign Assistance Programs; to Amend the Foreign Assistance Act of 1961, the Arms Export Control Act, and Other Acts to Authorize Appropriations for the Fiscal Year 1985 for International Security and Development Assistance, the Peace Corps and the International Development Association; and for Other Purposes*, S. Rep. No. 98-400 (1984).

US Congress. Senate. *The International Security Assistance and Arms Export Control Act of 1977. Report from the Committee on Foreign Relations*, S. Rep. No. 95-195 (1977).

US Congress. Senate. *The Mutual Security Act of 1959. Report of the Committee on Foreign Relations on S. 1451, Together with Minority Views*, S. Rep. No. 86-412 (1959).

#### ***PRESIDENTIAL DOCUMENTS***

Carter, Jimmy. "Presidential Directive NSC-30; Subject: Human Rights," February 17, 1978, Federation of American Scientists Intelligence Resource Program, *www.fas.org/irp/offdocs/pd/index.html*.

Eisenhower, Dwight D. "Remarks at the People-to-People Conference—September 11, 1956." *Public Papers of the Presidents of the United States* 1956 (1956), 749.

———. "Farewell Radio and Television Address to the American People—January 17, 1961." *Public Papers of the Presidents of the United States* 1960–61 (1961 1960), 1035–40.

———. "Report to Congress on the Mutual Security Program for the First Half of Fiscal Year 1960" (May 2, 1960).

Kennedy, John F. *Final Annual Report on the Operations of the Mutual Security Program*. H.R. Doc. No. 87-432 (June 12, 1962).

———. "Letter to the President of the Senate and to the Speaker of the House Transmitting Bill Implementing the Message on Foreign Aid—May 26, 1961." *Public Papers of the Presidents of the United States* 1961 (1961), 407.

———. "National Security Action Memorandum No. 88; Memorandum to the Secretary of Defense; Subject: Training for Latin American Armed Forces," September 5, 1961, Federation of American Scientists Intelligence Resource Program, *www.fas.org/irp/offdocs/nsam-jfk/index.html.*

———. "National Security Action Memorandum No. 89; Memorandum to the Secretary of State; Subject: Soviet Training of Ghanaian Troops," September 5, 1961, Federation of American Scientists Intelligence Resource Program, *www.fas.org/irp/offdocs/nsam-jfk/index.html.*

———. "National Security Action Memorandum No. 119; Memorandum to the Secretary of State and Secretary of Defense; Subject: Civic Action," December 18, 1961, Federation of American Scientists Intelligence Resource Program, *www.fas.org/irp/offdocs/nsam-jfk/index.html.*

———. *Report to the Congress on the Foreign Assistance Program for the Fiscal Year 1962.* Serial Set Vol. No. 12611, Session Vol. No. 47. Washington, DC: 88th Cong., 1st Sess., August 28, 1963.

———. "Special Message to the Congress on Foreign Aid—March 22, 1961." *Public Papers of the Presidents of the United States* 1961 (March 22, 1961), 203–13.

———. "Statement by the President on Ordering Resumption of Underground Nuclear Tests—September 5, 1961." *Public Papers of the Presidents of the United States* (September 5, 1961), 589.

———. "Statement by the President upon Signing the Foreign Assistance Act—September 4, 1961." *Public Papers of the Presidents of the United States* (September 4, 1961), 588.

———. "Urgent National Needs: Address of the President of the United States," 87th Cong. 1st Sess., Doc. No. 174 (May 25, 1961).

Truman, Harry S. "Inaugural Address—Thursday, January 20, 1949." *Inaugural Addresses of the Presidents of the United States* 1989 (1989), 285–92.

### OTHER US GOVERNMENT SOURCES

Amos, Harry O., George R. Bieber, John R. Johnson Jr., and Charles S. T. Mallett. *US Training of Foreign Military Personnel, Volume 1: Executive Summary.* McLean, VA: General Research Corporation, Tactical Warfare Operations (March 1979).

———. *US Training of Foreign Military Personnel, Volume 2: Main Report.* McLean, VA: General Research Corporation, Tactical Warfare Operations (March 1979).

Berry, Laverle Bennette. *Ghana: A Country Study, Area Handbook Series.* Washington, DC: Library of Congress Federal Research Division, 1995.

General Accounting Office. *School of the Americas: US Military Training for Latin American Countries; Report to the Ranking Minority Member, Committee on National Security, House of Representatives*. Washington, DC: General Accounting Office, National Security and International Affairs Division (August 22, 1996).

———. *Security Assistance: Observations on the International Military Education and Training Program*. Washington, DC: General Accounting Office, National Security and International Affairs Division (June 14, 1990).

Gilpatrick, Roswell. "Memorandum for the Service Secretaries from the Deputy Secretary of Defense, September 13, 1963; Subject: An Informational Program for Foreign Military Trainees and Visitors in the United States." Memo. Washington, DC, September 13, 1963. FOIA Request, in author's possession.

Grimmett, Richard F. *An Overview of United States Military Assistance Programs*. CRS Report for Congress. Washington, DC: Congressional Research Service, March 29, 1988.

———. *Foreign Assistance and Related Programs Appropriations for 1986*. Washington, DC: US Government Printing Office, 1985.

Grimmett, Richard F., and Mark P. Sullivan. *US Army School of the Americas: Background and Congressional Concerns*. CRS Report for Congress. Washington, DC: Congressional Research Service, 2001.

Manolas, Spiro C., and Louis J. Samelson. "The United States International Military Education and Training (IMET) Program: A Report to Congress." *DISAM Journal of International Security Assistance Management* 12, no. 3 (Spring 1990): 12–47.

McCoy, William H. *Senegal and Liberia: Case Studies in US IMET Training and Its Role in Internal Defense and Development*. N-3637-USDP. Santa Monica, CA: RAND, 1994.

Peterson, Rudolph A. *US Foreign Assistance in the 1970s: A New Approach; Report to the President from the Task Force on International Development*. Washington, DC: US Government Printing Office (March 4, 1970).

Polk, Frank L. *Invitation to Latin American Republics to Send Students to Military and Naval Academies*. Washington, DC: US Government Printing Office, 1919.

Taw, Jennifer Morrison. *Thailand and the Philippines: Case Studies in US IMET Training and Its Role in Internal Defense and Development*. Santa Monica, CA: RAND, 1994.

———. *The Effectiveness of Training International Military Students in Internal Defense and Development, Executive Summary*. Santa Monica, CA: RAND, 1993.

Taw, Jennifer Morrison, and William H. McCoy. *International Military Student Training: Beyond Tactics*. A RAND Note. Santa Monica, CA: RAND, 1993.

US Army. *Fort Leavenworth: From Frontier Post to Home of the United States Army Command and General Staff College*, 1964.

US Department of Defense. "Department of Defense Directive 5410.17 'An Informational Program for Foreign Military Trainees and Visitors in the United States,'" January 15, 1965, Combined Arms Research Library.

———. "Department of Defense Directive 5410.17 'An Informational Program for Foreign Military Trainees and Visitors in the United States,'" March 1, 1985. *www.dtic.mil/dtic/tr/fulltext/u2/a272252.pdf.*

———. "Department of Defense Instruction 5410.17 United States Field Studies Program (FSP) for International Military and Civilian Students and Military-Sponsored Visitors" (September 15, 2006) *www.dtic.mil/whs/directives/corres/pdf/541017p.pdf.*

US President's Committee to Study the US Military Assistance Program (1958–1959). *Conclusions Concerning the Mutual Security Program. Communication from the President of the United States Transmitting the Final Report of the President's Committee to Study the United States Military Assistance Program, with the Several Studies Which Are Annexes Thereto. August 20, 1959.* Washington, DC: US Government Printing Office, 1959.

DSAA (Defense Security Assistance Agency). *Congressional Presentation, Security Assistance Programs,* FY 1979–1990.

## WEB-BASED SOURCES

Ameyibor, Francis. "Ghana Remembers Brigadier Joseph Michel's 50 Years Anniversary." Ghanaian News. *GhanaWeb*, accessed September 7, 2011, *www.ghanaweb.com/GhanaHomePage/features/artikel.php?ID=218069.*

"Our History," Draper Investment Company, accessed October 30, 2013, *www.draperco.com/english/our-history/.*

"Inflation Calculator," Bureau of Labor Statistics, accessed March 5, 2014, *www.bls.gov/data/inflation_calculator.htm.*

Library of the Friedrich Ebert Foundation home page, accessed September 10, 2013, *library.fes.de/index_en.htm.*

US Army Combined Arms Center and Fort Leavenworth public home page, accessed March 20, 2012, *usacac.army.mil/cac2/index.asp.*

## BOOKS AND ARTICLES

Abaka, Edmund. "Traders, Slaves, and Soldiers: The Hausa Diaspora in Ghana (Gold Coast and Asante) in the Nineteenth and Early Twentieth Centuries." In *Movements, Borders, and Identities in Africa*, edited by Toyin Falola and Aribidesi Adisa Usman, 185–99. Rochester Studies in African History and the Diaspora, vol. 40. Rochester, NY: University of Rochester Press, 2009.

Aboagye, Festus B. *The Ghana Army: A Concise Contemporary Guide to Its Centennial Regimental History, 1897–1999*. Accra, Ghana: Sedco, 1999.

Achebe, Chinua. "English and the African Writer." *Transition*, no. 18 (January 1, 1965): 27–30.

Ackah, William. "The Intersection of African Identities in the Twenty-First Century: Old and New Diasporas and the African Continent." In *Reframing Contemporary Africa: Politics, Economics, and Culture in the Global Era*, edited by Peyi Soyinka-Airewele and Rita Kiki Edozie, 131–40. Washington, DC: CQ Press, 2010.

"A Congo Martyr." *West Africa*, no. 2312 (September 23, 1961): 57.

Adas, Michael. *Dominance by Design: Technological Imperatives and America's Civilizing Mission*. Cambridge, MA: Belknap Press of Harvard University Press, 2006.

———. *Machines as the Measure of Men: Science, Technology, and Ideologies of Western Dominance*. Ithaca, NY: Cornell University Press, 1989.

Addae, S. Kojo. *A Short History of Ghana Armed Forces*. Accra: Ministry of Defence of Ghana Armed Forces, 2005.

———. *History of Ghana Airforce*. Vol. 4 of Military Histories. Accra, Ghana: Eureka Publications, 2012.

———. *History of Ghana Armed Forces: Military Organizations and Regiments*. Vol. 3 of Military Histories. Accra, Ghana: Eureka Publications, c2010.

Adekson, J. Bayo. "Pay, Promotion, and Other Self-Interests of Military Intervention in Politics." *Military Affairs* 45, no. 1 (February 1, 1981): 18–22.

Afrifa, Akwasi A. *The Ghana Coup, 24th February 1966*. London: Cass, 1966.

Akyeampong, Emmanuel. "Africans in the Diaspora: The Diaspora and Africa." *African Affairs* 99, no. 395 (April 1, 2000): 183–215.

———. "'Diasporas,' Mobility and the Social Imaginary: Getting Ahead in West Africa." *Journal of Third World Studies* 27, no. 1 (Spring 2010): 25–41.

Alexander, H. T. *African Tightrope: My Two Years as Nkrumah's Chief of Staff*. London: Pall Mall, 1965.

Altınay, Ayşe Gül. *The Myth of the Military Nation: Militarism, Gender, and Education in Turkey*. New York: Palgrave Macmillan, 2004.

Amoah, Michael. *Reconstructing the Nation in Africa: The Politics of Nationalism in Ghana*. International Library of African Studies 19. London: Palgrave Macmillan, 2007.

Anderson, Benedict. *Imagined Communities: Reflections on the Origin and Spread of Nationalism*. Revised Edition. London: Verso, 2006.

Anyidoho, Henry Kwami. *Guns Over Kigali*. Accra, Ghana: Woeli Publishing Services, 1997.

———. *My Journey . . . Every Step*. Accra, Ghana: Sub-Saharan Publishers, 2010.

Appiah, Kwame Anthony. *Cosmopolitanism: Ethics in a World of Strangers*. Issues of Our Time. New York: W. W. Norton, 2006.

———. *In My Father's House: Africa in the Philosophy of Culture*. New York: Oxford University Press, 1992.

Apter, David E. *Ghana in Transition*. 2nd Revised. Princeton, NJ: Princeton University Press, 1972.

———. "Nkrumah, Charisma, and the Coup." *Daedalus* 97, no. 3 (July 1, 1968): 757–92.

———. *The Gold Coast in Transition*. Princeton, NJ: Princeton University Press, 1955.

Arkhurst, Frederick S. "A Brief Survey of Disarmament Negotiations." In *Arms and African Development: Proceedings of the First Pan-African Citizens' Conference*, edited by Frederick S. Arkhurst, 3–21. Praeger Special Studies in International Economics and Development. New York: Praeger, 1972.

———, ed. *Arms and African Development: Proceedings of the First Pan-African Citizens' Conference*. Praeger Special Studies in International Economics and Development. New York: Praeger, 1972.

Arlinghaus, Bruce E. *Military Development in Africa: The Political and Economic Risks of Arms Transfers*. Boulder, CO: Westview Press, 1984.

Arndt, Richard T. *The First Resort of Kings: American Cultural Diplomacy in the Twentieth Century*. Dulles, VA: Potomac Books, 2005.

Arthur, John A. *African Diaspora Identities: Negotiating Culture in Transnational Migration*. Lanham, MD: Lexington Books, 2010.

Assensoh, A. B., and Yvette Alex-Assensoh. *African Military History and Politics: Ideological Coups and Incursions, 1900–Present*. New York: Palgrave Macmillan, 2002.

Austin, Dennis. "The Ghana Armed Forces and Ghanaian Society." *Third World Quarterly* 7, no. 1 (January 1, 1985): 90–101.

Austin, Dennis, and Robin Luckham, eds. *Politicians and Soldiers in Ghana, 1966–1972*. Studies in Commonwealth Politics and History 3. London: Cass, 1975.

Bâ, Mariama. *So Long a Letter*. African Writers Series 248. London: Heinemann, 1989.

Banani, Amin. *The Modernization of Iran*. Stanford, CA: Stanford University Press, 1961.

Bauer, Péter Tamás. *Dissent on Development; Studies and Debates in Development Economics*. Cambridge, MA: Harvard University Press, 1972.

Bayly, C. A. *The Birth of the Modern World: 1780–1914*. Malden, MA: Blackwell, 2004.

Baynham, Simon. "Divide et Impera: Civilian Control of the Military in Ghana's Second and Third Republics." *Journal of Modern African Studies* 23, no. 4 (December 1, 1985): 623–42.

———. "Introduction: Armed Forces in Africa." In *Military Power and Politics in Black Africa*, edited by Simon Baynham. London: Croom Helm, 1986.
———. *The Military and Politics in Nkrumah's Ghana*. Boulder, CO: Westview Press, 1988.
Beidler, Philip D. *The Good War's Greatest Hits: World War II and American Remembering*. Athens: University of Georgia Press, 1998.
Bender, Thomas. "Introduction: Historians, the Nation, and the Plenitude of Narratives." In *Rethinking American History in a Global Age*, edited by Thomas Bender. Berkeley: University of California Press, 2002.
Bennett, Valerie Plave. "Epilogue: Malcontents in Uniform." In *Politicians and Soldiers in Ghana, 1966–1972*, edited by Dennis Austin and Robin Luckham, 300–312. London: Cass, 1975.
———. "The Motivation for Military Intervention: The Case of Ghana." *Western Political Quarterly* 26, no. 4 (December 1, 1973): 659–74.
Berger, Morroe. *Military Elite and Social Change*. Princeton, NJ: Center for International Studies, Princeton University, 1960.
Bienen, Henry S. "African Militaries as Foreign Policy Actors." In *Arms and the African: Military Influences on Africa's International Relations*, edited by William J Foltz and Henry S. Bienen, 153–70. New Haven, CT: Yale University Press, 1985.
———. "Armed Forces and National Modernization: Continuing the Debate." *Comparative Politics* 16, no. 1 (October 1, 1983): 1–16.
———, ed. *The Military and Modernization*. Chicago: Aldine Atherton, 1971.
Bill, James A. "The Military and Modernization in the Middle East." *Comparative Politics* 2, no. 1 (October 1969): 41.
Biney, Ama. *The Political and Social Thought of Kwame Nkrumah*. New York: Palgrave Macmillan, 2011.
Blundy, David, and Andrew Lycett. *Qaddafi and the Libyan Revolution*. Boston: Little, Brown, 1987.
Borstelmann, Thomas. *Apartheid's Reluctant Uncle: The United States and Southern Africa in the Early Cold War*. New York: Oxford University Press, 1993.
———. *The 1970s: A New Global History from Civil Rights to Economic Inequality*. Princeton, NJ: Princeton University Press, 2012.
Braham, Peter. *Key Concepts in Sociology*. SAGE Key Concepts. London: SAGE, 2013.
Brands, H. W. "The Age of Vulnerability: Eisenhower and the National Insecurity State." *American Historical Review* 94, no. 4 (1989): 963–89.
Brown, Carolyn A. "A 'Man' in the Village Is a 'Boy' in the Workplace: Colonial Racism, Worker Militance, and Igbo Notions of Masculinity in the Nigerian Coal Industry, 1930–1945." In *Men and Masculinities in Modern Africa*,

edited by Lisa A. Lindsay and Stephan Miescher, 156–74. Portsmouth, NH: Heinemann, 2003.

Brown, Kevin K. "The Military and Social Change in Colonial Tanganyika: 1919–1964." PhD Dissertation, Michigan State University, 2001.

Brubaker, Rogers, and Frederick Cooper. "Beyond 'Identity.'" *Theory and Society* 29, no. 1 (February 1, 2000): 1–47.

Bull, Hedley. "The State's Positive Role in World Affairs." *Daedalus* 108, no. 4 (October 1, 1979): 111–23.

Burbank, Jane, and Frederick Cooper. *Empires in World History: Power and the Politics of Difference*. Princeton, NJ: Princeton University Press, 2010.

Burke, Timothy James. *Lifebuoy Men, Lux Women: Commodification, Consumption and Cleanliness in Modern Zimbabwe*. Durham, NC: Duke University Press, 1996.

Campbell, Horace. "Pan-Africanism and African Liberation." In *Imagining Home: Class, Culture, and Nationalism in the African Diaspora*, edited by Sidney J. Lemelle and Robin D. G. Kelley, 285–307. London: Verso, 1994.

Campbell, James T. *Middle Passages: African American Journeys to Africa, 1787–2005*. New York: Penguin Press, 2006.

Chakrabarty, Dipesh. "The Difference-Deferral of a Colonial Modernity." In *Tensions of Empire: Colonial Cultures in a Bourgeois World*, edited by Frederick Cooper and Ann Laura Stoler, 373–405. Berkeley: University of California Press, 1997.

Chazan, Naomi. *An Anatomy of Ghanaian Politics: Managing Political Recession, 1969–1982*. Boulder, CO: Westview Press, 1983.

Chomsky, Noam. *American Power and the New Mandarins*. New York: Pantheon Books, 1969.

Clarke, Kamari Maxine. "New Spheres of Transnational Formations: Mobilizations of Humanitarian Diasporas." *Transforming Anthropology* 18, no. 1 (2010): 48–65.

Cleveland, Harlan. *The Overseas Americans*. 1st Edition. The Carnegie Series in American Education. New York: McGraw-Hill, 1960.

Clune, John V. "The Abongo Abroad: Military Internationalism, Travel, Training, and Peace in Ghana and the United States, 1960–1992." PhD Dissertation, University of Kansas, 2014.

Cobbs, Elizabeth A. "Decolonization, the Cold War, and the Foreign Policy of the Peace Corps." *Diplomatic History* 20, no. 1 (1996): 79–105.

Coleman, James S., and Belmont Brice Jr. "The Role of the Military in Sub-Saharan Africa." In *The Role of the Military in Underdeveloped Countries*, edited by John J. Johnson, 359–405. Princeton, NJ: Princeton University Press, 1962.

Collins, Robert O., and James M. Burns. *A History of Sub-Saharan Africa*. 2nd Edition. New York: Cambridge University Press, 2013.
Cooper, Frederick. *Colonialism in Question: Theory, Knowledge, History*. Berkeley: University of California Press, 2005.
———. "Industrial Man Goes to Africa." In *Men and Masculinities in Modern Africa*, edited by Lisa A. Lindsay and Stephan Miescher, 128–38. Portsmouth, NH: Heinemann, 2003.
———. "The Dialectics of Decolonization: Nationalism and Labor Movements in Postwar French Africa." In *Tensions of Empire: Colonial Cultures in a Bourgeois World*, edited by Frederick Cooper and Ann Laura Stoler, 406–35. Berkeley: University of California Press, 1997.
Crocker, Chester A. "Military Dependence: The Colonial Legacy in Africa." *Journal of Modern African Studies* 12, no. 2 (June 1, 1974): 265–86.
Cronin, James. *The World the Cold War Made: Order, Chaos, and the Return of History*. New York: Routledge, 1996.
Cullather, Nick. "'Fuel for the Good Dragon': The United States and Industrial Policy in Taiwan, 1950–1965." *Diplomatic History* 20, no. 1 (1996): 1–25.
———. "Modernization Theory." In *Explaining the History of American Foreign Relations*, 212–20. Cambridge: Cambridge University Press, 2004.
———. "Research Note: Development? It's History." *Diplomatic History* 24, no. 4 (October 2000): 641–53.
———. "The Foreign Policy of the Calorie." *American Historical Review* 112, no. 2 (April 1, 2007): 337–64.
———. *The Hungry World: America's Cold War Battle Against Poverty in Asia*. Cambridge, MA: Harvard University Press, 2010.
Cunliffe, Philip. *Legions of Peace: UN Peacekeepers from the Global South*. London: Hurst, 2013.
Davin, Anna. "Imperialism and Motherhood." In *Tensions of Empire: Colonial Cultures in a Bourgeois World*, edited by Frederick Cooper and Ann Laura Stoler, 87–151. Berkeley: University of California Press, 1997.
Decalo, Samuel. *Coups and Army Rule in Africa: Studies in Military Style*. New Haven, CT: Yale University Press, 1976.
———. "Military Rule in Africa: Etiology and Morphology." In *Military Power and Politics in Black Africa*, edited by Simon Baynham, 38–66. London: Croom Helm, 1986.
———. "Modalities of Civil-Military Stability in Africa." *Journal of Modern African Studies* 27, no. 4 (December 1, 1989): 547–78.
Dodwell, Henry. *The Founder of Modern Egypt*. Cambridge: University Press, 1931.
Dowse, Robert. "Military and Police Rule." In *Politicians and Soldiers in Ghana, 1966–1972*, edited by Dennis Austin and Robin Luckham, 16–33. London: Cass, 1975.

Dudziak, Mary L. *Cold War Civil Rights: Race and the Image of American Democracy*. Princeton, NJ: Princeton University Press, 2000.

Echenberg, Myron J. *Colonial Conscripts: The Tirailleurs Sénégalais in French West Africa, 1857–1960*. Social History of Africa Series. Portsmouth, NH: Heinemann, 1991.

Edib, Halide. *Turkey Faces West: A Turkish View of Recent Changes and Their Origin*. New Haven, CT: Yale University Press, 1930.

Ekbladh, David. *The Great American Mission: Modernization and the Construction of an American World Order*. Princeton, NJ: Princeton University Press, 2009.

Endy, Christopher. "Travel and World Power: Americans in Europe, 1890–1917." *Diplomatic History* 22, no. 4 (2002): 565–94.

Enloe, Cynthia H. *Bananas, Beaches and Bases: Making Feminist Sense of International Politics*. Berkeley: University of California Press, 2000.

Erskine, Emmanuel A. *Mission with UNIFIL: An African Soldier's Reflections*. London: Hurst, 1989.

Fahmy, Khaled. *All the Pasha's Men: Mehmed Ali, His Army, and the Making of Modern Egypt*. Cambridge: Cambridge University Press, 1997.

Feit, Edward. "Pen, Sword, and People: Military Regimes in the Formation of Political Institutions." *World Politics* 25, no. 2 (January 1, 1973): 251–73.

Field, Thomas C. "Ideology as Strategy: Military-Led Modernization and the Origins of the Alliance for Progress in Bolivia." *Diplomatic History* 36, no. 1 (January 1, 2012): 147–83.

Finnemore, Martha. "Redefining Development at the World Bank." In *International Development and the Social Sciences: Essays on the History and Politics of Knowledge*, edited by Frederick Cooper and Randall M. Packard, 203–27. Berkeley: University of California Press, 1997.

First, Ruth. *Power in Africa*. 1st American Edition. New York: Pantheon Books, 1970.

Fitch, John S. "The Political Impact of US Military Aid to Latin America: Institutional and Individual Effects." *Armed Forces and Society* 5, no. 3 (Spring 1979): 360–86.

Frank, Andre Gunder. "The Development of Underdevelopment." *Monthly Review* 18, no. 4 (1966): 17–31.

Fraser, Donald M. "Human Rights and US Foreign Policy: Some Basic Questions Regarding Principles and Practice." *International Studies Quarterly* 23, no. 2 (June 1, 1979): 174–85.

———. "New Directions in Foreign Aid." *World Affairs* 129, no. 4 (January 1, 1967): 244–50.

Frimpong, Brig. Gen. Daniel Kwadjo. *Leadership and the Challenges of Command: The Ghana Military Academy Experience*. Accra: Afram Publications, 2003.

Fry, Joseph A. "Place Matters: Domestic Regionalism and the Formation of American Foreign Policy." *Diplomatic History* 36, no. 3 (2012): 451–82.

Gaines, Kevin Kelly. *American Africans in Ghana: Black Expatriates and the Civil Rights Era*. Chapel Hill: University of North Carolina Press, 2006.

Gill, Lesley. *The School of the Americas: Military Training and Political Violence in the Americas*. American Encounters/Global Interactions Series. Durham, NC: Duke University Press, 2004.

Gilman, Nils. *Mandarins of the Future: Modernization Theory in Cold War America*. Baltimore: Johns Hopkins University Press, 2003.

Gilroy, Paul. *The Black Atlantic: Modernity and Double Consciousness*. Cambridge, MA: Harvard University Press, 1993.

Gingyera-Pincywa, A. G. G., and Ali A. Mazrui. "Regional Development and Regional Disarmament: Some African Perspectives." In *Arms and African Development: Proceedings of the First Pan-African Citizens' Conference*, edited by Frederick S. Arkhurst, 31–45. Praeger Special Studies in International Economics and Development. New York: Praeger, 1972.

Gorman, Daniel. *The Emergence of International Society in the 1920s*. Cambridge: Cambridge University Press, 2012.

Gravino, Katharine Swift, David R. Segal, and Mady Wechsler Segal. "Lightfighters' Wives." In *Peacekeepers and Their Wives: American Participation in the Multinational Force and Observers*, edited by David R. Segal and Mady Wechsler Segal, 140–56. Contributions in Military Studies, no. 147. Westport, CT: Greenwood Press, 1993.

Gutteridge, William F. *Military Institutions and Power in the New States*. New York: Praeger, 1965.

———. "The Political Role of African Armed Forces: The Impact of Foreign Military Assistance." *African Affairs* 66, no. 263 (April 1, 1967): 93–103.

Gyau-Boakye, P. "Environmental Impacts of the Akosombo Dam and Effects of Climate Change on the Lake Levels." *Environment, Development and Sustainability* 3, no. 1 (2001): 17.

Gyimah-Boadi, E., and Donald Rothchild. "Rawlings, Populism, and the Civil Liberties Tradition in Ghana." *Issue: A Journal of Opinion* 12, no. 3/4 (October 1, 1982): 64–69.

Halperin, Morton H. "The Gaither Committee and the Policy Process." *World Politics* 13, no. 3 (April 1, 1961): 360–84.

Hannah, John A. "New Directions in Foreign Aid for the 1970's." *American Journal of Agricultural Economics* 52, no. 2 (May 1, 1970): 302–7.

Hannerz, Ulf. "Cosmopolitans and Locals in World Culture." In *Global Culture: Nationalism, Globalization and Modernity: A Theory Culture and Society Special Issue*, edited by Mike Featherstone, 237–94. London: SAGE, 1990.

Hansen, Emmanuel. *Ghana Under Rawlings: Early Years*. Lagos, Nigeria: Malthouse Press, 1991.

Hansen, Emmanuel, and Paul Collins. "The Army, the State, and the 'Rawlings Revolution' in Ghana." *African Affairs* 79, no. 314 (January 1, 1980): 3–23.

Harris, Lillian Craig. *Libya: Qadhafi's Revolution and the Modern State*. Boulder, CO: Westview Press, 1986.

Harshe, Rajen. "Africa: Trends in US Imperialism." *Social Scientist* 12, no. 11 (November 1, 1984): 19–33.

Hartman, Saidiya V. *Lose Your Mother: A Journey Along the Atlantic Slave Route*. New York: Farrar, Straus and Giroux, 2007.

Hayter, Teresa. *Aid as Imperialism*. Baltimore: Penguin, 1971.

*Hearts and Minds*. Directed by Peter Davis. BBS Productions, 1974. DVD. Criterion Collection, 2002.

Henk, Dan. *Uncharted Paths, Uncertain Vision US Military Involvements in Sub-Saharan Africa in the Wake of the Cold War*. Occasional Paper 18. Colorado Springs, CO: USAF Institute for National Security Studies, US Air Force Academy, 1998.

Hoffman, Elizabeth Cobbs. *All You Need Is Love: The Peace Corps and the Spirit of the 1960s*. Cambridge, MA: Harvard University Press, 1998.

———. "Diplomatic History and the Meaning of Life: Toward a Global American History." *Diplomatic History* 21, no. 4 (1997): 499–518.

Holden, Robert H. "The Real Diplomacy of Violence: United States Military Power in Central America, 1950–1990." *International History Review* 15, no. 2 (1993): 283–322.

Hopkins, Keith. "Civil-Military Relations in Developing Countries." *British Journal of Sociology* 17, no. 2 (June 1, 1966): 165–82.

Horne, Gerald. "Race from Power: US Foreign Policy and the General Crisis of 'White Supremacy.'" *Diplomatic History* 23, no. 3 (1999): 437–61.

Howe, Herbert M. *Ambiguous Order: Military Forces in African States*. Boulder, CO: Lynne Rienner Publishers, 2001.

Hunt, G. D. "Recollections of the Canadian Armed Forces Training Team in Ghana, 1961–1968." *Canadian Defense Quarterly* 18, no. 5 (April 1989): 43–50.

Hunt, Michael. *Ideology and US Foreign Policy*. New Haven, CT: Yale University Press, 2009.

Huntington, Samuel P. *Changing Patterns of Military Politics*. Vol. 3 of International Yearbook of Political Behavior Research. New York: Free Press of Glencoe, 1962.

———. "Political Development and Political Decay." *World Politics* 17, no. 3 (April 1, 1965): 386–430.

———. *Political Order in Changing Societies*. New Haven, CT: Yale University Press, 1973.

———. *The Soldier and the State: The Theory and Politics of Civil-Military Relations*. Cambridge, MA: Belknap Press of Harvard University Press, 1957.

Hutchful, Eboe. "A Tale of Two Regimes: Imperialism, the Military and Class in Ghana." *Review of African Political Economy*, no. 14 (January 1, 1979): 36–55.

———. "Institutional Decomposition and Junior Ranks' Political Action in Ghana." In *The Military and Militarism in Africa*, edited by Eboe Hutchful and Abdoulaye Bathily, 211–56. Dakar: Codesria, 1998.

———. "Military Policy and Reform in Ghana." *Journal of Modern African Studies* 35, no. 2 (June 1, 1997): 251–78.

———. "Reconstructing Civil-Military Relations and the Collapse of Democracy in Ghana, 1979-81." *African Affairs* 96, no. 385 (October 1, 1997): 535–60.

Im, Hyug Baeg. "The Rise of Bureaucratic Authoritarianism in South Korea." *World Politics* 39, no. 2 (January 1, 1987): 231–57.

Iriye, Akira. *Cultural Internationalism and World Order*. Baltimore: Johns Hopkins University Press, 1997.

———. *Global Community the Role of International Organizations in the Making of the Contemporary World*. Berkeley: University of California Press, 2002.

———. "Internationalizing International History." In *Rethinking American History in a Global Age*, edited by Thomas Bender, 47–62. Berkeley: University of California Press, 2002.

Irwin, Ryan M. *Gordian Knot: Apartheid and the Unmaking of the Liberal World Order*. Oxford Studies in International History. Oxford: Oxford University Press, 2012.

Janowitz, Morris. *The Military in the Political Development of New Nations: An Essay in Comparative Analysis*. Chicago: University of Chicago Press, 1964.

Johnson, John J., ed. *The Role of the Military in Underdeveloped Countries*. Princeton, NJ: Princeton University Press, 1962.

Karnow, Stanley. *In Our Image: America's Empire in the Philippines*. New York: Random House, 1989.

Kennan, George F. *American Diplomacy*. Chicago: University of Chicago Press, 1984.

Kilford, Christopher R. *The Other Cold War: Canada's Military Assistance to the Developing World, 1945–1975*. Kingston, Ontario: Canadian Defence Academy Press, 2010.

Killingray, David. "Imagined Martial Communities: Recruiting for the Military and Police in Colonial Ghana, 1860–1960." In *Ethnicity in Ghana: The Limits of Invention*, edited by Carola Lentz and Paul Nugent, 119–36. New York: St. Martin's Press, 2000.

———. "Soldiers, Ex-Servicemen, and Politics in the Gold Coast, 1939–50." *Journal of Modern African Studies* 21, no. 03 (1983): 523–34.

Kolko, Gabriel. *Confronting the Third World: United States Foreign Policy, 1945–1980*. New York: Pantheon Books, 1988.

Kolko, Joyce, and Gabriel Kolko. *The Limits of Power: The World and United States Foreign Policy, 1945–1954*. New York: Harper & Row, 1972.

Korang, Kwaku Larbi. *Writing Ghana, Imagining Africa: Nation and African Modernity*. Rochester Studies in African History and the Diaspora. Rochester, NY: University of Rochester Press, 2003.

Kraus, Jon. "Arms and Politics in Ghana." In *Soldier and State in Africa: A Comparative Analysis of Military Intervention and Political Change*, edited by Claude E. Welch, 154–223. Evanston, IL: Northwestern University Press, 1970.

Kroes, Rob. "American Empire and Cultural Imperialism: A View from the Receiving End." *Diplomatic History* 23, no. 3 (1999): 463–77.

Latham, Michael E. *Modernization as Ideology: American Social Science and "Nation Building" in the Kennedy Era*. Chapel Hill: University of North Carolina Press, 2000.

Lauren, Paul Gordon, Gordon A. Craig, and Alexander L. George. *Force and Statecraft: Diplomatic Challenges of Our Time*. 4th Edition. Oxford: Oxford University Press, USA, 2006.

Lefever, Ernest W. "Arms Transfers, Military Training, and Domestic Politics." In *Arms Transfers in the Modern World*, edited by Stephanie G. Neuman and Robert E. Harkavy, 276–93. New York: Praeger, 1979.

———. *Spear and Scepter; Army, Police, and Politics in Tropical Africa*. Washington, DC: Brookings Institution, 1970.

———. "The Military Assistance Training Program." *Annals of the American Academy of Political and Social Science* 424 (March 1, 1976): 85–95.

Leffler, Melvyn P. *A Preponderance of Power: National Security, the Truman Administration and the Cold War*. Stanford, CA: Stanford University Press, 1992.

———. *For the Soul of Mankind: The United States, the Soviet Union, and the Cold War*. New York: Hill and Wang, 2007.

Lerner, Daniel. *The Passing of Traditional Society: Modernizing the Middle East*. Glencoe, IL: Free Press, 1958.

Lerner, Daniel, and Richard D. Robinson. "Swords and Ploughshares: The Turkish Army as a Modernizing Force." *World Politics* 13, no. 1 (October 1, 1960): 19–44.

Lovell, John P., and C. I. Eugene Kim. "The Military and Political Change in Asia." *Pacific Affairs* 40, no. 1/2 (April 1, 1967): 113–23.

Lundestad, Geir. "'Empire by Invitation' in the American Century." *Diplomatic History* 23, no. 2 (1999): 189–217.

Lutz, Catherine. *Homefront: A Military City and the American Twentieth Century*. Boston: Beacon Press, 2001.

MacGonagle, Elizabeth. *Crafting Identity in Zimbabwe and Mozambique*. Rochester, NY: University of Rochester Press, 2007.

Mahoney, Richard D. *JFK: Ordeal in Africa*. New York: Oxford University Press, 1983.

Manela, Erez. *The Wilsonian Moment: Self-Determination and the International Origins of Anticolonial Nationalism*. Oxford: Oxford University Press, 2009.

Manger, Leif O., and Munzoul A. M. Assal. "Diasporas Within and Without Africa—Dynamism, Heterogeneity, Variation." In *Diasporas Within and Without Africa: Dynamism, Heterogeneity, Variation*, edited by Lief O. Manger and Munzoul A. M. Assal, 7–31. Uppsala, Sweden: Nordiska Afrikainstitutet, 2006.

Mann, Gregory. *Native Sons: West African Veterans and France in the Twentieth Century*. Durham, NC: Duke University Press, 2006.

Manning, Patrick. *The African Diaspora: A History through Culture*. New York: Columbia University Press, 2009.

Manuh, Takyiwaa. "'Efie' or the Meanings of 'Home' Among Female and Male Ghanaian Migrants in Toronto, Canada and Returned Migrants to Ghana." In *New African Diasporas*, edited by Khalid Koser, 140–59. London: Routledge, 2003.

Martin, Michel L. "Of Arms and the Man: A Short Intellectual History of Morris Janowitz's Contribution to the Sociology of the Military." In *The Military, Militarism, and the Polity: Essays in Honor of Morris Janowitz*, edited by Michel L. Martin and Ellen Stern McCrate, 1–31. New York: Free Press, 1984.

Mayer, Philip, and Iona Mayer. *Townsmen or Tribesmen; Conservatism and the Process of Urbanization in a South African City*. Xhosa in Town 2. Cape Town: Oxford University Press, 1961.

Mazower, Mark. "An International Civilization? Empire, Internationalism and the Crisis of the Mid-Twentieth Century." *International Affairs* 82, no. 3 (May 1, 2006): 553–66.

———. *Governing the World: The History of an Idea*. New York: Penguin Press, 2012.

Mazrui, Ali A. "Soldiers as Traditionalizers: Military Rule and the Re-Africanization of Africa." *World Politics* 28, no. 2 (January 1, 1976): 246–72.

Mazrui, Ali A., and Michael Tidy. *Nationalism and New States in Africa: From about 1935 to the Present*. London: Heinemann, 1984.

McCormick, Thomas J. *America's Half-Century: United States Foreign Policy in the Cold War and After*. Baltimore: Johns Hopkins University Press, 1995.

McCoy, Alfred W. *Closer Than Brothers: Manhood at the Philippine Military Academy*. New Haven, CT: Yale University Press, 1999.

Meriwether, James H. "'Worth a Lot of Negro Votes': Black Voters, Africa, and the 1960 Presidential Campaign." *Journal of American History* 95, no. 3 (December 1, 2008): 737–63.

Millikan, Max F., and Walt W. Rostow. *A Proposal: Key to an Effective Foreign Policy.* New York: Harper & Brothers, 1957.

Milne, David. *America's Rasputin: Walt Rostow and the Vietnam War.* New York: Hill and Wang, 2008.

Moore, Raymond A. "The Peaceful Uses of Military Forces in Underdeveloped Areas: A Review Essay." *Journal of Developing Areas* 4, no. 1 (October 1, 1969): 112–19.

Moskos, Charles C. *Peace Soldiers: The Sociology of a United Nations Military Force.* Chicago: University of Chicago Press, 1976.

Mower, A. Glenn. *Human Rights and American Foreign Policy: The Carter and Reagan Experiences.* Studies in Human Rights, no. 7. New York: Greenwood Press, 1987.

Moya, Jose C. *Cousins and Strangers: Spanish Immigrants in Buenos Aires, 1850–1930.* Berkeley: University of California Press, 1998.

Moynihan, Daniel P. "The United States in Opposition." *Commentary* 59, no. 3 (March 1975): 31–45.

Muehlenbeck, Philip E. *Betting on the Africans: John F. Kennedy's Courting of African Nationalist Leaders.* Oxford: Oxford University Press, 2012.

———. "Kennedy and Touré: A Success in Personal Diplomacy." *Diplomacy and Statecraft* 19 (2008): 69–95.

Nasser, Gamal Abdel. "The Egyptian Revolution." *Foreign Affairs* 33, no. 2 (January 1, 1955): 199–211.

Navari, Cornelia. *Internationalism and the State in the Twentieth Century.* London: Routledge, 2000.

Nelson-Pallmeyer, Jack. *School of Assassins: Guns, Greed, and Globalization.* Revised Edition. Maryknoll, NY: Orbis Books, 2001.

Ngugi wa Thiongo. *Decolonising the Mind: The Politics of Language in African Literature.* London: J. Currey, 1986.

Niblock, Timothy. "A New Political System in Sudan." *African Affairs* 73, no. 293 (October 1974): 408–18.

Ninkovich, Frank A. *The Wilsonian Century: US Foreign Policy Since 1900.* Chicago: University of Chicago Press, 1999.

Nkrumah, Kwame. *Africa Must Unite.* New Edition. New York: International Publishers, 1970.

———. *Class Struggle in Africa.* 1st Edition. New York: International Publishers, 1970.

———. *Dark Days in Ghana.* London: Panaf, 1968.

———. *I Speak of Freedom: A Statement of African Ideology*. London: Heinemann, 1961.

———. *Revolutionary Path*. London: Panaf, 1973.

———. *The Autobiography of Kwame Nkrumah*. London: Panaf, 1973.

Ocran, A. K. *Politics of the Sword: A Personal Memoir on Military Involvement in Ghana and of Problems of Military Government*. London: Rex Collings, 1977.

O'Donnell, Guillermo A. *Modernization and Bureaucratic-Authoritarianism: Studies in South American Politics*. Politics of Modernization Series. Berkeley: Institute of International Studies, University of California, 1979.

Oduro-Kwarteng, Steve. *The Memoirs of a Colonel Retired*. Accra, Ghana: self-published, 2009.

Omitoogun, Wuyi. *Military Expenditure Data in Africa: A Survey of Cameroon, Ethiopia, Ghana, Kenya, Nigeria and Uganda*. Stockholm International Peace Research Institute (SIPRI) Reports 17. Oxford: Oxford University Press, 2003.

Osgood, Kenneth Alan. *Total Cold War: Eisenhower's Secret Propaganda Battle at Home and Abroad*. Lawrence: University of Kansas Press, 2006.

Owusu, Stephen. *Writing for Peace*. Ghana: tecnic Channel, 2009.

Owusu, Thomas Y. "The Role of Ghanaian Immigrant Associations in Toronto, Canada." *International Migration Review* 34, no. 4 (December 1, 2000): 1155–81.

Pach, Chester J. *Arming the Free World: The Origins of the United States Military Assistance Program, 1945–1950*. Chapel Hill: University of North Carolina Press, 1991.

Packenham, Robert A. *Liberal America and the Third World: Political Development Ideas in Foreign Aid and Social Science*. Princeton, NJ: Princeton University Press, 1973.

Painter, David S. "Explaining US Relations with the Third World." *Diplomatic History* 19, no. 3 (Summer 1995): 525.

Parker, Jason C. "'Made-in-America Revolutions'? The 'Black University' and the American Role in the Decolonization of the Black Atlantic." *Journal of American History* 96, no. 3 (December 1, 2009): 727–50.

Parsons, Timothy. *The African Rank-and-File: Social Implications of Colonial Military Service in the King's African Rifles, 1902–1964*. Social History of Africa Series. Portsmouth, NH: Heinemann, 1999.

Pool, Ithiel de Sola. "The Necessity for Social Scientists Doing Research for Governments." *Background* 10, no. 2 (August 1, 1966): 111–22.

Price, Robert M. "A Theoretical Approach to Military Rule in New States: Reference-Group Theory and the Ghanaian Case." *World Politics* 23, no. 3 (April 1, 1971): 399–430.

———. "Military Officers and Political Leadership: The Ghanaian Case." *Comparative Politics* 3, no. 3 (April 1, 1971): 361–79.

Pye, Lucian W. "Armies in the Process of Political Modernization." In *The Role of the Military in Underdeveloped Countries*, edited by John J. Johnson, 69–90. Princeton, NJ: Princeton University Press, 1962.

———. *Guerrilla Communism in Malaya: Its Social and Political Meaning.* Princeton, NJ: Princeton University Press, 1956.

Rakove, Robert B. *Kennedy, Johnson, and the Nonaligned World.* Cambridge: Cambridge University Press, 2013.

Reddy, Chandan. "Modern." In *Keywords for American Cultural Studies*, edited by Bruce Burgett and Glenn Hendler, 160–64. New York: New York University Press, 2007.

Reichley, John. *International Officers: A Century of Participation at the United States Army Command and General Staff College.* Leavenworth, KS: US Army Command and General Staff College, 1994.

Research and Development Department, ed. *Ghana Armed Forces Staff College: An Historical Perspective.* Accra, Ghana: Jamieson's Cambridge Faxbooks, 1996.

Rodgers, Daniel T. *Atlantic Crossings: Social Politics in a Progressive Age.* Cambridge, MA: Belknap Press of Harvard University Press, 1998.

Rosenberg, Emily S. *Spreading the American Dream.* New York: Hill and Wang, 1982.

Rosh, Robert M. "Third World Militarization: Security Webs and the States They Ensnare." *Journal of Conflict Resolution* 32, no. 4 (December 1, 1988): 671–98.

Rostow, Walt W. *The Diffusion of Power: An Essay in Recent History.* New York: Macmillan, 1972.

———. *The Process of Economic Growth: A Non-Communist Manifesto.* 2d Edition. Oxford: Clarendon Press, 1960.

———. *The Stages of Economic Growth: A Non-Communist Manifesto.* Cambridge: University Press, 1960.

———. "Vietnam and Asia." *Diplomatic History* 20, no. 3 (1996): 467–71.

Rustow, Dankwart A. "The Army and the Founding of the Turkish Republic." *World Politics* 11, no. 4 (July 1, 1959): 513–52.

Ruttan, Vernon W. *United States Development Assistance Policy: The Domestic Politics of Foreign Economic Aid.* Johns Hopkins Studies in Development Series. Baltimore: Johns Hopkins University Press, 1996.

Ryan, Selwyn. "The Theory and Practice of African One Partyism: The CPP Re-examined." *Canadian Journal of African Studies / Revue Canadienne Des Études Africaines* 4, no. 2 (April 1, 1970): 145–72.

Sacca, John Wands. "Like Strangers in a Foreign Land: Chinese Officers Prepared at American Military Colleges, 1904–37." *Journal of Military History* 70, no. 3 (July 1, 2006): 703–42.

Savage, Jesse Dillon, and Jonathan D. Caverley. "Foreign Military Training and Coup Propensity." Unpublished article, October 1, 2015. *www.jonathancaverley.com/uploads/2/9/7/2/29726853/caverleysavage_sept17_2015.pdf.*

Seigel, Micol. "Beyond Compare: Comparative Method after the Transnational Turn." *Radical History Review* 2005, no. 91 (December 21, 2005): 62–90.

———. *Uneven Encounters: Making Race and Nation in Brazil and the United States*. American Encounters/Global Interactions Series. Durham, NC: Duke University Press, 2009.

Shils, Edward. "The Military in the Political Development of the New States." In *The Role of the Military in Underdeveloped Countries*, edited by John J. Johnson, 7–67. Princeton, NJ: Princeton University Press, 1962.

Siekmann, Robert C. R. *National Contingents in United Nations Peace-Keeping Forces*. Dordrecht, Netherlands: Martinus Nijhoff, 1991.

Simpson, Bradley R. *Economists with Guns: Authoritarian Development and US-Indonesian Relations, 1960–1968*. Stanford, CA: Stanford University Press, 2008.

Skrbiš, Zlatko, and Ian Woodward. *Cosmopolitanism: Uses of the Idea*. London: SAGE, 2013.

Smaldone, Joseph P. "US Arms Transfers and Security Assistance Programs in Africa: A Review and Policy Perspective." In *Arms for Africa: Military Assistance and Foreign Policy in the Developing World*, edited by Bruce E. Arlinghaus, 179–220. Lexington, MA: D. C. Heath, 1983.

Snead, David L. *The Gaither Committee, Eisenhower, and the Cold War*. Columbus: Ohio State University Press, 1999.

Snyder, Sarah B. *Human Rights Activism and the End of the Cold War: A Transnational History of the Helsinki Network*. Human Rights in History. New York: Cambridge University Press, 2011.

Suri, Jeremi. *Power and Protest: Global Revolution and the Rise of Detente*. Cambridge, MA: Harvard University Press, 2003.

Taffet, Jeffrey. *Foreign Aid as Foreign Policy: The Alliance for Progress in Latin America*. New York: Routledge, 2007.

Thomas, Daniel C. "Boomerangs and Superpowers: International Norms, Transnational Networks and US Foreign Policy." *Cambridge Review of International Affairs* 15, no. 1 (April 2002): 25–44.

Thomas, Lewis V., and Richard Nelson Frye. *The United States and Turkey and Iran*. American Foreign Policy Library. Cambridge: Harvard University Press, 1951.

Thompson, Robert Grainger Ker. *Defeating Communist Insurgency: Experiences from Malaya and Vietnam*. Studies in International Security, no. 10. London: Chatto & Windus, 1966.

Tiger, Lionel. "Nkrumah's Ghana and the Theory of Charisma." *Bulletin of African Studies in Canada / Bulletin Des Études Africaines Au Canada* 2, no. 1 (November 1, 1964): 2–10.

Toynbee, Arnold. "The Modernisation of the Middle East." *Journal of the Royal Institute of International Affairs* 8, no. 4 (July 1, 1929): 344–66.

Tyrrell, Ian. *Reforming the World: The Creation of America's Moral Empire*. Princeton, NJ: Princeton University Press, 2010.

United Nations Information Center. *Ghana: Four Decades of UN and Regional Peacekeeping*. Accra, Ghana: United Nations Information Center, 2000.

Vagts, Alfred. *A History of Militarism*. Revised Edition. London: Hollis and Carter, 1959.

Von Eschen, Penny M. *Race Against Empire: Black Americans and Anticolonialism, 1937–1957*. Ithaca, NY: Cornell University Press, 1997.

———. *Satchmo Blows up the World: Jazz Ambassadors Play the Cold War*. Cambridge, MA: Harvard University Press, 2004.

Wagnleitner, Reinhold. "The Empire of the Fun, or Talkin' Soviet Union Blues: The Sound of Freedom and US Cultural Hegemony in Europe." *Diplomatic History* 23, no. 3 (1999): 499–524.

Wallerstein, Immanuel Maurice. "Modernization: Requiescat in Pace." In *The Capitalist World-Economy: Essays*, 132–37. Studies in Modern Capitalism. Cambridge: Cambridge University Press, 1979.

Welch, Claude E. Jr. "The African Military and Political Development." In *The Military and Modernization*, edited by Henry Bienen, 212–33. Chicago: Aldine Atherton, 1971.

———. "Civilian Control of the Military: Myth and Reality." In *Civilian Control of the Military: Theory and Cases from Developing Countries*, edited by Claude E. Welch Jr., 1–42. Albany: State University of New York Press, 1976.

———, ed. *Soldier and State in Africa: A Comparative Analysis of Military Intervention and Political Change*. Evanston, IL: Northwestern University Press, 1970.

Weschler, Lawrence. "A Reporter at Large: The Great Exception I—Liberty." *New Yorker*, April 3, 1989, 43–85.

Westad, Odd Arne. *The Global Cold War: Third World Interventions and the Making of Our Times*. Cambridge: Cambridge University Press, 2007.

Wicker, Tom. "A Theological War." *Diplomatic History* 20, no. 3 (1996): 445–49.

Wiley, David. "Militarizing Africa and African Studies and the US Africanist Response." *African Studies Review* 55, no. 2 (2012): 147–61.

Williams, William Appleman. *The Tragedy of American Diplomacy*. New York: Norton, 1988.

Windle, Charles, and T. R. Vallance. "Optimizing Military Assistance Training." *World Politics* 15, no. 1 (October 1, 1962): 91–107.

Young, Robert. *Postcolonialism: A Very Short Introduction*. Oxford: Oxford University Press, 2003.

Zeiler, Thomas W. "Just Do It! Globalization for Diplomatic Historians." *Diplomatic History* 25, no. 4 (2001): 529–51.

Zeleza, Paul Tiyambe. "Reconceptualizing African Diasporas: Notes from a Historian." *Transforming Anthropology* 18, no. 1 (2010): 74–78.
Zimmerman, Andrew. *Alabama in Africa: Booker T. Washington, the German Empire, and the Globalization of the New South*. Princeton, NJ: Princeton University Press, 2010.
Zolberg, Aristide R. "The Structure of Political Conflict in the New States of Tropical Africa." *American Political Science Review* 62, no. 1 (March 1, 1968): 70–87.

# Index

Numbers in **bold** refer to tables.